Selected Titles in This Series

665 **George Lawrence Ashline,** The defect relation of meromorphic maps on parabolic manifolds, 1999

664 **Xia Chen,** Limit theorems for functionals of ergodic Markov chains with general state space, 1999

663 **Ola Bratteli and Palle E. T. Jorgensen,** Iterated function systems and permutation representation of the Cuntz algebra, 1999

662 **B. H. Bowditch,** Treelike structures arising from continua and convergence groups, 1999

661 **J. P. C. Greenlees,** Rational S^1-equivariant stable homotopy theory, 1999

660 **Dale E. Alspach,** Tensor products and independent sums of $\mathcal{L}_p$-spaces, $1 < p < \infty$, 1999

659 **R. D. Nussbaum and S. M. Verduyn Lunel,** Generalizations of the Perron-Frobenius theorem for nonlinear maps, 1999

658 **Hasna Riahi,** Study of the critical points at infinity arising from the failure of the Palais-Smale condition for n-body type problems, 1999

657 **Richard F. Bass and Krzysztof Burdzy,** Cutting Brownian paths, 1999

656 **W. G. Bade, H. G. Dales, and Z. A. Lykova,** Algebraic and strong splittings of extensions of Banach algebras, 1999

655 **Yuval Z. Flicker,** Matching of orbital integrals on $GL(4)$ and $GSp(2)$, 1999

654 **Wancheng Sheng and Tong Zhang,** The Riemann problem for the transportation equations in gas dynamics, 1999

653 **L. C. Evans and W. Gangbo,** Differential equations methods for the Monge-Kantorovich mass transfer problem, 1999

652 **Arne Meurman and Mirko Primc,** Annihilating fields of standard modules of $\mathfrak{sl}(2,\mathbb{C})^\sim$ and combinatorial identities, 1999

651 **Lindsay N. Childs, Cornelius Greither, David J. Moss, Jim Sauerberg, and Karl Zimmermann,** Hopf algebras, polynomial formal groups, and Raynaud orders, 1998

650 **Ian M. Musson and Michel Van den Bergh,** Invariants under Tori of rings of differential operators and related topics, 1998

649 **Bernd Stellmacher and Franz Georg Timmesfeld,** Rank 3 amalgams, 1998

648 **Raúl E. Curto and Lawrence A. Fialkow,** Flat extensions of positive moment matrices: Recursively generated relations, 1998

647 **Wenxian Shen and Yingfei Yi,** Almost automorphic and almost periodic dynamics in skew-product semiflows, 1998

646 **Russell Johnson and Mahesh Nerurkar,** Controllability, stabilization, and the regulator problem for random differential systems, 1998

645 **Peter W. Bates, Kening Lu, and Chongchun Zeng,** Existence and persistence of invariant manifolds for semiflows in Banach space, 1998

644 **Michael David Weiner,** Bosonic construction of vertex operator para-algebras from symplectic affine Kac-Moody algebras, 1998

643 **Józef Dodziuk and Jay Jorgenson,** Spectral asymptotics on degenerating hyperbolic 3-manifolds, 1998

642 **Chu Wenchang,** Basic almost-poised hypergeometric series, 1998

641 **W. Bulla, F. Gesztesy, H. Holden, and G. Teschl,** Algebro-geometric quasi-periodic finite-gap solutions of the Toda and Kac-van Moerbeke hierarchies, 1998

640 **Xingde Dai and David R. Larson,** Wandering vectors for unitary systems and orthogonal wavelets, 1998

639 **Joan C. Artés, Robert E. Kooij, and Jaume Llibre,** Structurally stable quadratic vector fields, 1998

638 **Gunnar Fløystad,** Higher initial ideals of homogeneous ideals, 1998

637 **Thomáš Gedeon,** Cyclic feedback systems, 1998

(*Continued in the back of this publication*)

The Defect Relation of Meromorphic Maps on Parabolic Manifolds

MEMOIRS
of the
American Mathematical Society

Number 665

The Defect Relation of Meromorphic Maps on Parabolic Manifolds

George Lawrence Ashline

May 1999 • Volume 139 • Number 665 (fourth of 5 numbers) • ISSN 0065-9266

American Mathematical Society
Providence, Rhode Island

1991 *Mathematics Subject Classification.*
Primary 32–XX.

Library of Congress Cataloging-in-Publication Data

Ashline, George Lawrence, 1967–
The defect relation of meromorphic maps on parabolic manifolds / George Lawrence Ashline.
p. cm. — (Memoirs of the American Mathematical Society, ISSN 0065-9266 ; no. 665)
"May 1999, volume 139, number 665 (fourth of 5 numbers)."
Includes bibliographical references.
ISBN 0-8218-1069-3 (alk. paper)
1. Value distribution theory. 2. Functions, Meromorphic. 3. Mappings (Mathematics) I. Title. II. Series.
QA3.A57 no. 665
[QA331.7]
510 s—dc21
[515′.982] 99-19211
CIP

Memoirs of the American Mathematical Society

This journal is devoted entirely to research in pure and applied mathematics.

Subscription information. The 1999 subscription begins with volume 137 and consists of six mailings, each containing one or more numbers. Subscription prices for 1999 are \$448 list, \$358 institutional member. A late charge of 10% of the subscription price will be imposed on orders received from nonmembers after January 1 of the subscription year. Subscribers outside the United States and India must pay a postage surcharge of \$30; subscribers in India must pay a postage surcharge of \$43. Expedited delivery to destinations in North America \$35; elsewhere \$130. Each number may be ordered separately; *please specify number* when ordering an individual number. For prices and titles of recently released numbers, see the New Publications sections of the *Notices of the American Mathematical Society.*

Back number information. For back issues see the *AMS Catalog of Publications.*

Subscriptions and orders should be addressed to the American Mathematical Society, P. O. Box 5904, Boston, MA 02206-5904. *All orders must be accompanied by payment.* Other correspondence should be addressed to Box 6248, Providence, RI 02940-6248.

Memoirs of the American Mathematical Society is published bimonthly (each volume consisting usually of more than one number) by the American Mathematical Society at 201 Charles Street, Providence, RI 02904-2294. Periodicals postage paid at Providence, RI. Postmaster: Send address changes to Memoirs, American Mathematical Society, P. O. Box 6248, Providence, RI 02940-6248.

This publication is indexed in *Science Citation Index*®, *SciSearch*®, *Research Alert*®, *CompuMath Citation Index*®, *Current Contents*®/*Physical, Chemical & Earth Sciences.*
Printed in the United States of America.

∞ The paper used in this book is acid-free and falls within the guidelines established to ensure permanence and durability.
Visit the AMS home page at URL: http://www.ams.org/

10 9 8 7 6 5 4 3 2 1   04 03 02 01 00 99

Contents

Abstract

Suppose that M is a parabolic manifold of dimension m and V is a vector space of dimension $n+1>1$. Let $f : M \to \mathbb{P}(V)$ be a meromorphic map and $\mathfrak{G}$ be a finite set of meromorphic maps $g : M \to \mathbb{P}(V^*)$ in general position. Also, suppose that each g in $\mathfrak{G}$ grows slower than f. Then, the following defect relation can be proven:

$$\sum_{g\in\mathfrak{G}} \delta(f,g) \quad \leq \quad n+1.$$

This is derived using the original method of Cartan [3] and following the ideas of the Lemma of the Logarithmic Derivative over $\mathbb{C}^m$ given by Vitter [17] in 1977. In order to achieve this result, the mild assumption must be made that there exists a holomorphic m form Θ on M. This is equivalent to assuming that there exists a nontrivial holomorphic section of the canonical bundle K of M. Also, the Ricci curvature of the manifold M and the norm of Θ taken along the fibres of K play key roles in the derivation the Lemma of the Logarithmic Derivative in this case.

This result holds on all parabolic, Stein manifolds. In particular, if M_i is a noncompact, parabolic Riemann surface for $1 \leq i \leq m$, then this relation is valid on $M_1 \times ... \times M_m$. Also, if $\pi : M \to N$ is a holomorphic line bundle where N a connected, compact complex manifold of dimension $m-1>0$, then M is another example of a parabolic manifold on which the defect relation holds under the standard assumptions.

Notation

(1) $\mathbb{N}$ = the positive integers
(2) $\mathbb{Z}$ = the integers
(3) $\mathbb{Q}$ = the rational numbers
(4) $\mathbb{R}$ = the real numbers
(5) $\mathbb{C}$ = the complex numbers
(6) $\overline{\overline{\mathbb{R}}} = \mathbb{R} \cup \{-\infty, +\infty\}$
(7) $\overline{\overline{\mathbb{C}}} = \mathbb{C} \cup \{\infty\} = \mathbb{P}_1$, the Riemann sphere

If $p \in \mathbb{N}$ and S is a set, then

(8) $S^p = S \times ... \times S$ (p-times)
(9) $\#S$ = card S, the number of elements of S

If $a \in \overline{\overline{\mathbb{R}}}, b \in \overline{\overline{\mathbb{R}}}$, and $T \subseteq \overline{\overline{\mathbb{R}}}$, then

(10) *i.* $T[a,b] = \{x \in T | a \leq x \leq b\}$
ii. $T(a,b] = \{x \in T | a < x \leq b\}$
iii. $T[a,b) = \{x \in T | a \leq x < b\}$
iv. $T(a,b) = \{x \in T | a < x < b\}$

(11) *i.* $T^+ = T(0, +\infty]$
ii. $T_+ = T[0, +\infty]$
iii. $T^- = T[-\infty, 0)$
iv. $T_- = T[-\infty, 0]$

For $\rho \in \mathbb{Z}_+$,

$$(12)\quad i_\rho = \left(\frac{i}{2\pi}\right)^\rho (-1)^{\frac{\rho(\rho-1)}{2}} \rho!$$

If M is a complex manifold of dimension m, then

(13) *i.* $\mathfrak{T}(M)$ = the holomorphic tangent bundle of M
ii. $\mathfrak{T}^*(M)$ = the holomorphic cotangent bundle of M
iii. $\overline{\mathfrak{T}}(M)$ = the conjugate of $\mathfrak{T}(M)$
iv. $\overline{\mathfrak{T}}^*(M)$ = the conjugate of $\mathfrak{T}^*(M)$
v. $\bigwedge_m \mathfrak{T}^*(M)$ = the canonical bundle of M, denoted K_M or K

Introduction

Suppose M is a connected, complex manifold of dimension m. Suppose that there exists a parabolic exhaustion τ on M. Then, (M, τ) is called a parabolic manifold. Let V be a vector space of dimension $n+1 > 1$. Let $f : M \to \mathbb{P}(V)$ be a meromorphic map and $\mathfrak{G}$ a finite set of target meromorphic maps $g : M \to \mathbb{P}(V^*)$ in general position. Let T_f and T_g be the characteristic functions of f and g, respectively. Assume that each g in $\mathfrak{G}$ grows slower than f, i.e. that for $0 < s < r$,

$$\lim_{r\to\infty} \frac{T_g(r,s)}{T_f(r,s)} = 0.$$

Denote by $N_{f,g}$ the valence function of the divisor $\mu_{f,g}$. For f or g nonconstant and $0 < s < r$, the Nevanlinna defect of f for g is given by

$$0 \leq \delta(f,g) = 1 - \limsup_{r\to\infty} \frac{N_{f,g}(r,s)}{T_f(r,s) + T_g(r,s)} \leq 1.$$

Now, suppose $M = \mathbb{C}$ and $n = 1$. For q in $\mathbb{N}$, the incidence can be considered of the image of f and distinct fixed targets $\{a_j\}_{j=1}^q$ contained in $\mathbb{P}_1 = \mathbb{C} \cup \{\infty\}$. In this case, R. Nevanlinna derived in 1924 his defect relation

$$\sum_{j=1}^{q} \delta(f, a_j) \leq 2.$$

Notice that this gives a generalization of Picard's Theorem. Then, in 1929, Nevanlinna [5] conjectured that this defect relation remains valid if the fixed targets are replaced by a finite set $\mathfrak{G}$ of meromorphic target functions $g : \mathbb{C} \to \mathbb{P}_1$ growing slower than f. He proved this conjecture when $\mathfrak{G}$ consists of three target functions. C. T. Chuang [4] verified the conjecture for entire functions in 1964. In 1986, N. Steinmetz [8] generalized Chuang's method and proved Nevanlinna's conjecture.

Next, suppose that $n > 1$. Also, suppose that $f : \mathbb{C} \to \mathbb{P}(V)$ is a holomorphic curve and $\mathfrak{G}$ is a finite set of holomorphic curves $g : \mathbb{C} \to \mathbb{P}(V^*)$. Then, assuming each g in $\mathfrak{G}$ grows slower than f, W. Stoll [14] proved in 1986 that

$$\sum_{g\in\mathfrak{G}} \delta(f,g) \leq n(n+1).$$

In 1990, under the same assumptions, M. Ru-W. Stoll [7] proved that

$$\sum_{g\in\mathfrak{G}} \delta(f,g) \leq n+1,$$

where $n+1$ is a sharp bound. This result can be extended to meromorphic maps $f : \mathbb{C}^m \to \mathbb{P}_n$. Also in 1990, E. Bardis [2] further extended this result to ramified

Received by the editor June 6, 1996, and in revised form January 5, 1998 parabolic covering spaces M of $\mathbb{C}^m$. If $f : M \to \mathbb{P}(V)$ is a meromorphic map and $\mathfrak{G}$ a finite set of meromorphic maps $g : M \to \mathbb{P}(V^*)$, then Bardis [2] showed that

$$\sum_{g\in\mathfrak{G}} \delta(f,g) \quad \leq \quad n+1.$$

Now, consider M a parabolic manifold and $f : M \to \mathbb{P}(V)$ a meromorphic map. Let $\mathfrak{G}$ be a finite set of meromorphic maps $g : M \to \mathbb{P}(V^*)$ given as before. Then, it will be proven that

$$\sum_{g\in\mathfrak{G}} \delta(f,g) \quad \leq \quad n+1.$$

The proof of this Defect Relation relies heavily on the Lemma of the Logarithmic Derivative for meromorphic maps. It is also necessary to assume that there exists a holomorphic m form $\Theta \not\equiv 0$ on M. This result is part of my dissertation, which was written at the University of Notre Dame under the direction of Professor Wilhelm Stoll. I wish to thank Professor Stoll for his assistance with these efforts.

This paper contains four chapters leading up to this proof. Throughout this paper the notation used in Stoll [10] will be adopted. In the first section of the first chapter, some fundamental definitions and results are given. Then in the second section, a brief overview of Jacobian sections is presented. In the final section, the holomorphic forms Θ and $\mathbb{B}$ on M of degrees m and $m-1$, respectively, are introduced and the meromorphic differential operator D is defined.

In the second chapter, several different cases of the Lemma of the Logarithmic Derivative are presented. The chapter ends with a derivation of the Second Main Theorem for Fixed Targets.

In the third chapter, some of the theory is outlined that was developed by Ru-Stoll [7] and considered locally by Bardis [2]. With this theory, several important vector spaces are constructed and the Steinmetz map is introduced.

To begin the fourth and final chapter, a set of critical assumptions are made. In the first section, several important results are described which yield the Lemma of the Logarithmic Derivative with respect to this theory. In the second section, the Main Identity is derived. In the third section, estimates are found for the functions θ_α and Λ which are essential in the proof of the Second Main Theorem for Moving Targets. In the fourth section, the Product to Sum Estimates are stated which precede the culmination of this paper in the proof of the Defect Relation in the final section.

Chapter 1

1.1 Parabolic manifolds and the Ricci functions

The defect relation for moving targets has been sought after since the late 1920's. Chuang [4] provided a first step in this direction by finding a defect relation for entire holomorphic functions of a single complex variable. Good results have been obtained utilizing the methods of Steinmetz [8] and Ru-Stoll [7] which follow the proofs of the defect relations of Nevanlinna [5] and Cartan [3] in the fixed target case. The defect relations proven by Stoll using the Ahlfors-Weyl theory require undesirable assumptions in the moving target case. The Steinmetz technique as well as the Cartan proof require a Lemma of the Logarithmic Derivative which was proven for $\mathbb{C}$ by Nevanlinna [5], for $\mathbb{C}^n$ by Vitter [17], and for parabolic covering manifolds by Bardis [2]. Now, a Lemma of the Logarithmic Derivative on a parabolic manifold M is proven under the mild assumption that the canonical bundle K_M of M has a non-trivial holomorphic section. This lemma is then used to prove the defect relation for moving targets on M.

To begin with, assume that M is a connected, complex manifold of dimension m. The notion of a parabolic exhaustion must be introduced. A proper, C^∞ map $\tau : M \to \mathbb{R}_+$ is called an exhaustion of M.

For $r > 0$ and $S \subseteq M$, define

$$\begin{aligned}
\boldsymbol{v} &= dd^c\tau; \\
\omega &= dd^c \log\tau; \\
\sigma &= d^c \log\tau \wedge \omega^{m-1}; \\
S^+ &= \{x \in S | \boldsymbol{v}(x) > 0\}; \\
S_* &= \{x \in S | \tau(x) > 0\}; \\
S < r > &= \{x \in S | \tau(x) = r^2\}; \\
S(r) &= \{x \in S | \tau(x) < r^2\}; \\
S[r] &= \{x \in S | \tau(x) \le r^2\}.
\end{aligned}$$

If $\omega \geq 0$ and $\boldsymbol{v}^m \not\equiv 0 \equiv \omega^m = d\sigma$, then the exhaustion τ is called parabolic. As a result, $\boldsymbol{v} \geq 0$. If $\boldsymbol{v} > 0$, then τ is called strictly parabolic. If τ is a parabolic (resp. strictly parabolic) exhaustion for M, then (M, τ) is called a parabolic (resp. strictly parabolic) manifold.

Now, some basic properties and results of parabolic manifolds will be outlined. See Stoll [13], pp. 136-154 and Stoll [12], pp. 18-19 for these and more details.

First of all, suppose that X is a reduced, irreducible complex space of dimension m. Also, let $\Sigma(X)$ be the singular set of X and $\mathfrak{R}(X) = X - \Sigma(X)$ be the regular set. If G is a nonempty, open subset of X, then let dG denote the largest boundary submanifold of class C^∞ of $\mathfrak{R}(G)$ in $\mathfrak{R}(X)$. If dG is nonempty, then orient dG to the exterior of G. If $\overline{G}$ is compact, ∂G has locally finite Hausdorff measure of dimension $2m-1$, and $\partial G - dG$ has zero Hausdorff measure of dimension $2m-1$, then G is said to be a Stokes domain. If X is a form of class C^1 and degree $2m-1$ on X and if G is a Stokes domain in X, then Stokes theorem holds

$$\int_{dG} X = \int_G dX.$$

See Tung [16] for more details.

Define

$$\mathfrak{E}_\tau = \{r \in \mathbb{R}^+ | d\tau(x) \neq 0 \quad \forall x \in M < r >\}.$$

For r in $\mathfrak{E}_\tau$, $\partial M(r) = M < r >$ is a compact, real, $(2m-1)$-dimensional submanifold of class C^∞ of M oriented to the exterior of $M < r >$. Also, $\mathbb{R}^+ - \mathfrak{E}_\tau$ has measure zero. See p. 33 of Stoll [12]. Denote

$$\varsigma = \int_{M<r>} \sigma$$

which is positive and independent of r in $\mathfrak{E}_\tau$. In addition,

$$\int_{M[r]} \upsilon^m = \int_{M(r)} \upsilon^m = \varsigma r^{2m}.$$

Next, a divisor on M can be defined using its multiplicity function. If G is a nonempty, open subset of M, then suppose $f : G \to \mathbb{C}$ is a holomorphic function. Take x in G and suppose that $\alpha : U \to U'$ is a biholomorphic map of an open neighborhood U of x in G onto a ball U' in $\mathbb{C}^m$ such that $\alpha(x) = 0$. Then, there exists a unique sequence of homogeneous polynomials P_j defined on $\mathbb{C}^m$ such that $\deg P_j = j$ and

$$f|U = \sum_{j=0}^{\infty} P_j \circ \alpha.$$

Let $I = \{j \in \mathbb{Z}_+ | P_j \not\equiv 0\}$. Then, I is empty if and only if $f|U \equiv 0$. Define the zero multiplicity of f at x by

$$\mu_f^0(x) = \begin{cases} \min I & \text{if } I \neq \emptyset \\ \infty & \text{if } I = \emptyset. \end{cases}$$

The zero multiplicity is independent of the choice of α. A function $\nu : M \to \mathbb{Z}$ is called a divisor if for each point x in M there exists a connected, open neighborhood U of x and if there exist holomorphic functions $g \not\equiv 0 \not\equiv h$ on U with $\nu|U = \mu_g^0 - \mu_h^0$. If $\nu \equiv 0$, then ν is called the null divisor. The set $\mathfrak{D} = \mathfrak{D}(M)$ of all divisors on

M forms a module. For $0 \not\equiv \nu$ in $\mathfrak{D}$, its support $S = \text{supp}\nu$ is an analytic subset of M of pure dimension $m-1$; if $\nu \equiv 0$, then suppν is empty. A divisor ν is non-negative as a function if and only if for every x in M, there exists a connected open neighborhood U of x and a holomorphic function $g \not\equiv 0$ on U such that $\nu|U = \mu_g^0$.

Now, consider the case when (M, τ) is an m-dimensional complex parabolic manifold. Suppose $\nu : M \to \mathbb{Z}$ is a divisor such that $S = \text{supp}\nu$. For all $t > 0$, define the <u>counting function</u> n_ν of ν when $m > 1$ by

$$n_\nu(t) = t^{2-2m} \int_{S[t]} \nu \boldsymbol{v}^{m-1}$$

and when $m = 1$ by

$$n_\nu(t) = \sum_{x \in S[t]} \nu(x).$$

In turn, the limit $n_\nu(t) \to n_\nu(0)$ for $t \to 0$, $t > 0$ exists. If $m > 1$, then for all $t > 0$,

$$n_\nu(t) = \int_{S_*[t]} \nu \omega^{m-1} + n_\nu(0).$$

Next, for $0 < s < r$, define the <u>valence function</u> N_ν of ν by

$$N_\nu(r, s) = \int_s^r n_\nu(t) \frac{dt}{t}.$$

If $\nu \geq 0$, then $n_\nu \geq 0$ and $N_\nu(\cdot, s) \geq 0$ increase such that the limits

$$\lim_{r \to \infty} \frac{N_\nu(r, s)}{\log r} = \lim_{t \to \infty} n_\nu(t) = n_\nu(\infty) \leq \infty$$

exist. Furthermore, ν is said to have <u>affine growth</u> if $n_\nu(\infty) < \infty$.

If $f \not\equiv 0$ is a meromorphic function on M, then the zero divisor $\mu_f^0 \geq 0$ and the pole divisor $\mu_f^\infty \geq 0$ are defined. For each z in M, on a connected, open neighborhood U of z, there exist holomorphic functions $g \not\equiv 0$ and $h \not\equiv 0$ satisfying

i. $\dim g^{-1}(0) \cap h^{-1}(0) \leq m - 2$;

ii. $g = f \cdot h$ on U.

Here, $\mu_f^0|U = \mu_g^0$ and $\mu_f^\infty|U = \mu_h^0$. The <u>divisor</u> of f is given by $\mu_f = \mu_g^0 - \mu_h^0$. For $0 < s < r$ in $\mathfrak{C}_\tau$ with s in $\mathfrak{C}_\tau$, <u>Jensen's Formula</u> holds:

$$N_{\mu_f}(r, s) = \int_{M\langle r \rangle} \log |f| \sigma - \int_{M\langle s \rangle} \log |f| \sigma.$$

Now, the notion of a meromorphic map between complex manifolds must be considered. Let M and N be complex manifolds of pure dimension m and n, respectively. Let S be a <u>thin</u> analytic subset of M, meaning that $A = M - S$ is a dense subset of M. Also, let $f : A \to N$ be a holomorphic map with graph $\Gamma = \{(x, f(x)) | x \in A\}$. Let $\overline{\Gamma}$ be the closure of Γ in $M \times N$ and $\pi : \overline{\Gamma} \to M$ be its projection onto M. If $\overline{\Gamma}$ is analytic and π is proper, then f is called a meromorphic map on M. If f is meromorphic, then there exists a unique, smallest, closed subset I_f of S such that f continues to a holomorphic map $f : M - I_f \to N$. Here, I_f is said to be the indeterminacy of f with dimension at most $m - 2$.

Let V be a complex vector space with dimension $n+1 > 1$. Denote $V - \{0\}$ by V_*. Define the associated projective space to V by $\mathbb{P}(V) = \frac{V_*}{\mathbb{C}_*}$. Then, the quotient map is given by $\mathbb{P} : V_* \to \mathbb{P}(V)$.

Now, set $N = \mathbb{P}(V)$ and choose A as before. Suppose that $f : A \to \mathbb{P}(V)$ is a holomorphic map. Also, suppose that U is a nonempty, open, connected subset of M. A holomorphic map $\mathfrak{v} : U \to V$ is called a representation of f if $\mathfrak{v} \not\equiv 0$ and if $f(x) = \mathbb{P}(\mathfrak{v}(x))$ for all x in $U \cap A$ such that $\mathfrak{v}(x) \neq 0$. If a is in U, then $\mathfrak{v}$ is called a representation at a. The representation is called reduced if $\dim \mathfrak{v}^{-1}(0) \leq m - 2$. If $\mathfrak{v} : U \to V$ is a reduced representation, then $U \cap I_f = \mathfrak{v}^{-1}(0)$. Also, f is meromorphic if and only if there exists a representation at every point of M.

Next, $\mathfrak{v} : U \to V$ is said to be a meromorphic representation of a meromorphic map f if for all p in U, there exists an open, connected neighborhood $U_p \subseteq U$ of p, a holomorphic function $0 \not\equiv \lambda : U_p \to \mathbb{C}$, and a representation $\mathfrak{w}$ such that

$$\mathfrak{w} = \lambda \mathfrak{v} \quad \text{or} \quad \mathfrak{v} = \frac{\mathfrak{w}}{\lambda}.$$

Now, if $\mathfrak{v} : U \to V$ and $\mathfrak{w} : U \to V$ are two meromorphic representations of f, then there exists a meromorphic function $0 \not\equiv \eta$ on U such that

$$\mathfrak{v} = \eta \mathfrak{w}.$$

If $\mathfrak{v}$ is a representation and $\mathfrak{w}$ a reduced representation of f, then there exists a holomorphic function $0 \not\equiv \lambda$ on U such that

$$\mathfrak{v} = \lambda \mathfrak{w}.$$

If $\mathfrak{v}$ and $\mathfrak{w}$ are both reduced representations of f, then the above λ will be nowhere zero.

If ν is a divisor on N with support S and $f^{-1}(S)$ is thin, then f is called free for ν. The pullback divisor $f^*(\nu)$ is uniquely defined on M in the following way: For each x in $M - I_f$, there exist open, connected neighborhoods U of x in $M - I_f$ and W of $f(x)$ in N and holomorphic functions $g \not\equiv 0$ and $h \not\equiv 0$ on W such that

i. $f(U) \subseteq W$;

ii. $\operatorname{supp}\mu_g^0 \subseteq S$ and $\operatorname{supp}\mu_h^0 \subseteq S$;

iii. $\nu | W = \mu_g^0 - \mu_h^0$.

Then, $g \circ f \not\equiv 0 \not\equiv h \circ f$ and

$$f^*(\nu)|U = \mu^0_{g \circ f} - \mu^0_{h \circ f}.$$

Since the dimension of I_f is at most $m - 2$, then by Proposition 2.1 of Stoll [13], $f^*(\nu)$ can be extended from $M - I_f$ to M.

Now, the notion of a zero divisor on M will be introduced. Let $\mathfrak{w} : M \to V$ be a vector function on M such that $\mathfrak{w} \not\equiv 0$. For each p in M, take an open connected neighborhood U of p on which are defined $\mathfrak{v} : U \to V$ and $g : U \to \mathbb{C}$ satisfying the following properties:

1. $\mathfrak{v}$, g are holomorphic on U;

2. $\mathfrak{w}|U = \mathfrak{v} \cdot g$;

3. $\dim \mathfrak{v}^{-1}(0) \leq m - 2$.

Then, the zero divisor of $\mathfrak{w}$ is given by $\mu_{\mathfrak{w}}|U = \mu_g$.

To show this is well-defined, choose another open, connected neighborhood $\tilde{U}$ of p on which are defined $\tilde{\mathfrak{v}} : \tilde{U} \to V$ and $\tilde{g} : \tilde{U} \to \mathbb{C}$. Suppose that

1.′ $\tilde{\mathfrak{v}}$, $\tilde{g}$ are holomorphic on $\tilde{U}$;

2.′ $\mathfrak{w}|U = \tilde{\mathfrak{v}} \cdot \tilde{g}$;

3.′ $\dim \tilde{\mathfrak{v}}^{-1}(0) \leq m - 2$.

Suppose $\mathfrak{e}_0, ..., \mathfrak{e}_n$ is a basis of V. Let $\mathfrak{v} = v_0\mathfrak{e}_0 + ... + v_n\mathfrak{e}_n$ and $\tilde{\mathfrak{v}} = \tilde{v_0}\mathfrak{e}_0 + ... + \tilde{v_n}\mathfrak{e}_n$ be the representations of $\mathfrak{v}$ and $\tilde{\mathfrak{v}}$ in terms of their coordinate functions. From lines 2. and 2.′ above, $g\mathfrak{v} = \tilde{g}\tilde{\mathfrak{v}}$. Hence, by lines 3. and 3.′ above, there exists an integer j with $0 \leq j \leq n$ such that $v_j \not\equiv 0 \not\equiv \tilde{v}_j$. This implies that $gv_j = \tilde{g}\tilde{v}_j$ or $\frac{g}{\tilde{g}} = \frac{\tilde{v}_j}{v_j}$. By a symmetric argument, $\frac{\tilde{g}}{g} = \frac{v_j}{\tilde{v}_j}$. Consequently, there exists a holomorphic function $h : U \cap \tilde{U} \to \mathbb{C}_*$ with $\tilde{g} = h \cdot g$. Hence, $\mu_{\tilde{g}} = \mu_g$.

Next, consider as before V be a complex vector space of dimension $n+1$ and $N = \mathbb{P}(V)$. The p^{th}Grassmann cone $\tilde{G}_p(V)$ is analytic and given by

$$\tilde{G}_p(V) \quad = \quad \{\mathfrak{v}_0 \wedge ... \wedge \mathfrak{v}_p | \mathfrak{v}_j \in V\}.$$

Also, the p^{th}Grassmann manifold $G_p(V) = \mathbb{P}(\tilde{G}_p(V))$ is a smooth, connected, complex submanifold of $\mathbb{P}(\underset{p+1}{\wedge} V)$ such that the dimension of $G_p(V) = (p+1)(n-p)$. For $v = \mathbb{P}(\mathfrak{v})$ in $G_p(V)$ with $\mathfrak{v} = \mathfrak{v}_0 \wedge ... \wedge \mathfrak{v}_p$ nowhere zero,

$$\begin{aligned} E(v) = E(\mathfrak{v}) &= \{\mathfrak{x} \in V | \mathfrak{v} \wedge \mathfrak{x} = 0\} \\ &= \mathbb{C}\mathfrak{v}_0 + ... + \mathbb{C}\mathfrak{v}_p \end{aligned}$$

is a (p+1)-dimensional linear subspace of V. Furthermore, $\ddot{E}(V) = \mathbb{P}(E(V))$ is a p-dimensional complex projective plane in $\mathbb{P}(V)$, also referred to as a p-plane. In addition, for $w = \mathbb{P}(\mathfrak{w})$ in $\mathbb{P}(V^*)$, define

$$E[w] = E[\mathfrak{w}] = \{\mathfrak{x} \in V | < \mathfrak{x}, \mathfrak{w} >= 0\}$$

which yields an n-dimensional linear subspace of V, whereas $\ddot{E}[W] = \mathbb{P}(E[w])$ is an $(n-1)$-plane in $\mathbb{P}(V)$. In other words, $\ddot{E}[W]$ is a hyperplane in $\mathbb{P}(V)$. See Chapter 3, part A of Stoll [15] for these and other details.

Suppose f and g are meromorphic maps from M to N. The projective distance between f and g, denoted by $\Box f, g\Box$, will be defined. Let U be an open, connected subset of M. Let $\mathfrak{v} : U \to V$ be a reduced representation of f and $\mathfrak{w} : U \to V^*$ a reduced representation of g. Then,

$$\square f, g \square | U \quad = \quad \frac{| < \mathfrak{v}, \mathfrak{w} > |}{\| \mathfrak{v} \| \| \mathfrak{w} \|}.$$

Note that $0 \leq \square f, g \square \leq 1$. Now, (f, g) are called free if and only if one of the following equivalent conditions hold:

1. There exists $z_0 \in M - (I_f \cup I_g)$ such that $f(z_0) \notin \ddot{E}[g(z_0)]$;

2. There exists representations $\mathfrak{v} : U \to V$ of f and $\mathfrak{w} : U \to V^*$ of g

   such that $< \mathfrak{v}, \mathfrak{w} > \not\equiv 0$;

3. For every pair of representations $\mathfrak{v} : U \to V$ of f and $\mathfrak{w} : U \to V^*$

   of g, $< \mathfrak{v}, \mathfrak{w} > \not\equiv 0$;

4. $\square f, g \square \not\equiv 0$.

Suppose that (f, g) are free. Define the intersection divisor of f and g by

$$\mu_{f,g} \quad = \quad \mu_{< \mathfrak{v}, \mathfrak{w} >}.$$

This is well-defined. On an open subset U_λ of M, suppose that $\mathfrak{v}_\lambda$ is a reduced representation of f and $\mathfrak{w}_\lambda$ is a reduced representation of g. Then, as before, $\mu_{f,g} | U_\lambda = \mu_{< \mathfrak{v}_\lambda, \mathfrak{w}_\lambda >}$. Also, on $U_{\lambda\mu} = U_\lambda \cap U_\mu$, there exist nonzero holomorphic functions $g_{\mu\lambda}$ and $h_{\mu\lambda}$ such that

$$\begin{aligned} \mathfrak{v}_\mu &= g_{\mu\lambda} \mathfrak{v}_\lambda; \\ \mathfrak{w}_\mu &= h_{\mu\lambda} \mathfrak{w}_\lambda. \end{aligned}$$

Therefore, since $< \mathfrak{v}_\mu, \mathfrak{w}_\mu > = g_{\mu\lambda} h_{\mu\lambda} < \mathfrak{v}_\lambda, \mathfrak{w}_\lambda >$ and $\mu^0_{g_{\mu\lambda} h_{\mu\lambda}} = 0$, then $\mu^0_{< \mathfrak{v}_\mu, \mathfrak{w}_\mu >} = \mu^0_{< \mathfrak{v}_\lambda, \mathfrak{w}_\lambda >}$. In turn, the valence function for the intersection divisor of f and g is given by

$$N_{f,g}(r, s) \quad = \quad N_{\mu_{f,g}}(r, s).$$

Now, suppose that L is a holomorphic line bundle over N and κ is a hermitian metric along the fibers of L. Suppose α is a holomorphic section in L such that its zero set $Z(\alpha)$ is thin. In turn, the zero divisor μ_α of α is defined such that $\mathrm{supp}\mu_\alpha = Z(\alpha)$. Now, (f, α) is called free if $f^{-1}(Z(\alpha))$ is thin. Therefore, if (f, α) is free, then the intersection divisor can be defined by $\mu_{f,\alpha} = f^*(\mu_\alpha)$ because L is a line bundle. See p. 142 of Stoll [13]. Then, the Chern form of L with respect to κ on $N - Z(\alpha)$ is given by

$$c(L, \kappa) \quad = \quad -dd^c \log \| \alpha \|^2_\kappa .$$

The Chern form $c(L, \kappa)$ is of class C^∞ and bidegree $(1, 1)$ on M. Also, $c(L, \kappa)$ is independent of the choice of α. For $0 < t$ in $\mathbb{R}$, define the spherical image of f by

$$A_f(t; L, \kappa) \quad = \quad \frac{1}{t^{2m-2}} \int_{M[t]} f^*(c(L, \kappa)) \wedge \boldsymbol{\upsilon}^{m-1}$$

and for $0 < s < r$, define the characteristic function of f by

$$T_f(r, s; L, \kappa) = \int_s^r A_f(t; L, \kappa)\frac{dt}{t}.$$

Also, for the intersection divisor $\mu_{f,\alpha}$, the counting function and valence function are denoted by $n_f(t; \alpha, L) \geq 0$ and $N_f(r, s; \alpha, L) \geq 0$, respectively. For r in $\mathfrak{E}_\tau$, the compensation function is given by

$$m_f(r; \alpha, L, \kappa) = \int_{M\langle r\rangle} \log \frac{1}{\| \alpha \circ f \|_\kappa}\sigma.$$

Then, for $0 < s < r$ in $\mathfrak{E}_\tau$ and s in $\mathfrak{E}_\tau$, the First Main Theorem for Line Bundles states

$$T_f(r, s; L, \kappa) = N_f(r, s; \alpha, L) + m_f(r; \alpha, L, \kappa) - m_f(s; \alpha, L, \kappa). \tag{1.1}$$

Since T_f and N_f are continuous functions of r and s, $m_f(\cdot; \alpha, L, \kappa)$ can be extended continuously to $\mathbb{R}^+$ such that 1.1 will hold for all $0 < s < r$.

The tautological bundle $\pi : \mathcal{O}(-1) \to \mathbb{P}(V)$ and the Hopf σ-process $\sigma : \mathcal{O}(-1) \to V$ are given by

$$\mathcal{O}(-1) = \{(x, \mathfrak{x}) \in \mathbb{P}(V) \times V | \mathfrak{x} \in E(x)\}$$

where π and σ are the respective projections. From the inclusion map $\iota : \mathcal{O}(-1) \to \mathbb{P}(V) \times V$, $\mathcal{O}(-1)$ can be viewed as a subbundle of $\mathbb{P}(V) \times V$, the trivial bundle. For x in $\mathbb{P}(V)$, the fiber $\mathcal{O}(-1)_x = E(x)$ is a line and thus $\mathcal{O}(-1)$ is a line bundle. Its dual $H = \mathcal{O}(-1)^* = \mathcal{O}(1)$ is also a line bundle, called the hyperplane section bundle over $\mathbb{P}(V)$.

Denote by V_M the trivial bundle $M \times V$. There exists uniquely, up to isomorphism, a holomorphic line bundle L_f and a holomorphic section $F = F_f$ of $V_M \otimes L_f$ over M given by the following property:

> Suppose $\mathfrak{v} : U \to V$ is a reduced representation of f. Define a section $\tilde{\mathfrak{v}}$ of V_M over U by $\tilde{\mathfrak{v}}(x) = (x, \mathfrak{v})$. Then, there exists a section $\overset{\triangle}{\mathfrak{v}}$ of L_f over U such that $Z(\overset{\triangle}{\mathfrak{v}}) = 0$ and $F|U = \tilde{\mathfrak{v}} \otimes \overset{\triangle}{\mathfrak{v}}$.

If H is the hyperplane section bundle of $\mathbb{P}(V)$, then $L_f|(M - I_f)$ is isomorphic to $f^*(H)$ on $M - I_f$. Hence, L_f is called the hyperplane section bundle of f and F_f the representation section of f. Here, F_f substitutes as a global reduced representation of f.

Suppose $\mathfrak{i}$ is a hermitian metric on V. Then, $\mathfrak{i}$ induces a hermitian metric $\mathfrak{i}$ on the fibers on $\mathbb{P}(V) \times V$ which then restricts to a hermitian metric $\mathfrak{i}$ on the fibers of $\mathcal{O}(-1)$. This defines a dual metric $\mathfrak{i}^*$ on the fibers of H which can then be denoted $\mathfrak{i}$ if no confusion results. In turn, the Chern form of the hermitian line bundle $(H, \mathfrak{i})$ is called the

Fubini-Study form on $\mathbb{P}(V)$ which is positive and denoted by

$$\Omega = c(H, \mathfrak{i}).$$

As before, suppose that (M, τ) is a parabolic manifold with dimension m. Also, suppose that $f : M \to \mathbb{P}(V)$ and $g : M \to \mathbb{P}(V^*)$ are meromorphic maps. If $t > 0$, define the spherical image function of f by

$$A_f(t) \quad = \quad A_f(t, H, \mathfrak{i}) = \frac{1}{t^{2m-2}} \int_{M[\ell]} f^*(\Omega) \wedge \upsilon^{m-1} \geq 0.$$

Then, for $0 < s < r$, define the characteristic function of f by

$$T_f(r, s) \quad = \quad T_f(r, s, H, \mathfrak{i}) = \int_s^r A_f(t) \frac{dt}{t} \geq 0.$$

Notice that if f is constant, then $A_f(t) \equiv 0$ and therefore $T_f(r, s) \equiv 0$. For r in $\mathfrak{E}_\tau$ and (f, g) free, define the compensation function by

$$m_{f,g}(r) \quad = \quad \int_{M<r>} \log \frac{1}{\Box f, g \Box} \sigma.$$

For $0 < s < r$ in $\mathfrak{E}_\tau$ and s in $\mathfrak{E}_\tau$, the First Main Theorem for Moving Targets then states that

$$T_f(r, s) + T_g(r, s) \quad = \quad N_{f,g}(r, s) + m_{f,g}(r) - m_{f,g}(s). \tag{1.2}$$

Since T_f, T_g, and $N_{f,g}$ are all continuous in r and s, then $m_{f,g}$ will extend to a continuous function on $\mathbb{R}^+$ such that 1.2 holds for all $0 < s < r$ in $\mathbb{R}$ and s in $\mathbb{R}$.

Now, assume that either f or g is non-constant. To measure the intersection of $f(M)$ with $\ddot{E}[g(z)]$ for every z in M, define the Nevanlinna defect by

$$\delta(f, g) \quad = \quad \liminf_{r \to \infty} \frac{m_{f,g}(r)}{T_f(r, s) + T_g(r, s)}.$$

By 1.2,

$$\begin{aligned} 0 \leq \delta(f, g) \quad &= \quad \liminf_{r \to \infty} \frac{T_f(r, s) + T_g(r, s) - N_{f,g}(r, s)}{T_f(r, s) + T_g(r, s)} \\ &= \quad 1 - \limsup_{r \to \infty} \frac{N_{f,g}(r, s)}{T_f(r, s) + T_g(r, s)} \\ &\leq \quad 1. \end{aligned}$$

Note that if $g = a$ is a constant map, then $T_g(r, s) = T_a(r, s) \equiv 0$ and the defect becomes

$$\begin{aligned} \delta(f, a) \quad &= \quad \liminf_{r \to \infty} \frac{m_{f,a}(r)}{T_f(r, s)} \\ &= \quad 1 - \limsup_{r \to \infty} \frac{N_{f,a}(r, s)}{T_f(r, s)}. \end{aligned}$$

Next, the Ricci functions for the parabolic exhaustion τ must be introduced. To do this, some of the results given on p. 122 of Stoll [10] will be used.

Suppose $\mathfrak{U} = \{U_\lambda\}_{\lambda\in\Lambda}$ is a family of subsets of M. For each $\lambda = (\lambda_0, ..., \lambda_j)$ in Λ^{j+1}, define

$$\begin{aligned} U_\lambda &= U_{\lambda_0,...,\lambda_j} = U_{\lambda_0} \cap ... \cap U_{\lambda_j} \text{ and} \\ \Lambda[j] &= \{\lambda \in \Lambda^{j+1} | U_\lambda \neq \emptyset.\} \end{aligned}$$

Then $\Lambda[j] = \Lambda[j, \mathfrak{U}]$ is called the jth nerve of $\mathfrak{U}$. If U_λ is open for all λ in Λ, then the family $\mathfrak{U}$ is called open. $\mathfrak{U}$ is called a covering of M if and only if $M = \bigcup_{\lambda\in\Lambda} U_\lambda$.

Let $\{U_\lambda, \mathfrak{z}_\lambda\}_{\lambda\in\Lambda}$ be a family of holomorphic charts on M such that $\mathfrak{U} = \{U_\lambda\}_{\lambda\in\Lambda}$ is an open covering of M and such that

$$\mathfrak{z}_\lambda = (z_\lambda^1, ..., z_\lambda^m) : U_\lambda \to U'_\lambda$$

is a holomorphic chart of M. Then, $\zeta_\lambda = dz_\lambda^1 \wedge ... \wedge dz_\lambda^m$ is a holomorphic form of bidegree $(m, 0)$ without zeros on U_λ.

For (λ, μ) in $\Lambda[1]$, there is a unique holomorphic function $\Delta_{\lambda\mu} : U_{\lambda\mu} \to \mathbb{C}_*$ such that $\zeta_\lambda = \Delta_{\lambda\mu}\zeta_\mu$ on $U_{\lambda\mu}$. Hence, $\{\Delta_{\lambda\mu}\}_{(\lambda,\mu)\in\Lambda[1]}$ will form the basic cocycle of the canonical bundle K of M. This cocycle satisfies the following conditions:

$$\begin{aligned} &i. & \Delta_{\lambda\lambda} &= 1 \text{ on } U_\lambda\,; \\ &ii. & \Delta_{\lambda\mu}\Delta_{\mu\lambda} &= 1 \text{ on } U_\lambda \cap U_\mu\,; \\ &iii. & \Delta_{\lambda\mu}\Delta_{\mu\rho}\Delta_{\rho\lambda} &= 1 \text{ on } U_\lambda \cap U_\mu \cap U_\rho. \end{aligned}$$

Let Ψ be a positive form of class C^∞ and degree $2m$ on M. For every λ in Λ, there exists a unique positive function Ψ_λ defined on U_λ such that

$$\Psi|U_\lambda = \Psi_\lambda i_m d\zeta_\lambda \wedge \overline{d\zeta_\lambda}. \tag{1.3}$$

Also, for (λ, μ) in $\Lambda[1]$,

$$\begin{aligned} \Psi_\mu i_m d\zeta_\mu \wedge \overline{d\zeta_\mu}\, |U_\lambda \cap U_\mu &= \Psi_\lambda i_m d\zeta_\lambda \wedge \overline{d\zeta_\lambda}\, |U_\lambda \cap U_\mu \\ &= \Psi_\lambda i_m \Delta_{\lambda\mu} d\zeta_\mu \wedge \overline{(\Delta_{\lambda\mu} d\zeta_\mu)}\, |U_\lambda \cap U_\mu \\ &= \Psi_\lambda i_m \left|\Delta_{\lambda\mu}\right|^2 d\zeta_\mu \wedge \overline{d\zeta_\mu}\, |U_\lambda \cap U_\mu\,. \end{aligned}$$

Thus, $\Psi_\mu = \Psi_\lambda \left|\Delta_{\lambda\mu}\right|^2$.

Since $dd^c \log |\Delta_{\lambda\mu}|^2 \equiv 0$, define the Ricci form of Ψ, denoted $Ric\ \Psi$, to be the unique form of bidegree $(1, 1)$ and class C^∞ on M such that

$$Ric\ \Psi|U_\lambda = dd^c \log \Psi_\lambda$$

for each λ in Λ. For all forms ψ and χ of bidegree $(m, 0)$ on M, define a hermitian metric κ_Ψ along the fibres of K by

$$\kappa_\Psi(\psi|\chi)\Psi = i_m \psi \wedge \overline{\chi}.$$

Abbreviate κ_Ψ by κ. Thus, $\kappa(d\zeta_\lambda|d\zeta_\lambda)\Psi = \| d\zeta_\lambda \|_\kappa^2\ \Psi = i_m d\zeta_\lambda \wedge \overline{d\zeta_\lambda}$. By 1.3, $\| d\zeta_\lambda \|_\kappa^2\ \Psi_\lambda = 1$ or $\Psi_\lambda = \frac{1}{\|d\zeta_\lambda\|_\kappa^2}$. Consequently, $Ric\ \Psi|U_\lambda = dd^c \log \frac{1}{\|d\zeta_\lambda\|_\kappa^2} = -dd^c \log \| d\zeta_\lambda \|_\kappa^2 = c(K, \kappa)$, where $c(K, \kappa)$ is the Chern form of the canonical bundle K for the hermitian metric κ.

Now, for $0 < s < r$, define the Ricci function of Ψ to be

$$Ric(r, s, \Psi) = \int_s^r \int_{M[t]} (Ric\ \Psi) \wedge \boldsymbol{v}^{m-1} t^{1-2m} dt. \tag{1.4}$$

Since $Ric\ \Psi = c(K, \kappa)$, then $Ric(r, s, \Psi) = T(r, s, K, \kappa)$, the characteristic function of the canonical bundle K with hermitian metric κ.

If $\Psi > 0$ is chosen to be any form of degree $2m$ and class C^∞ on M, then define a nonnegative, continuous function v on M such that $\boldsymbol{v}^m = v^2\Psi$. Note that v^2 is of class C^∞ on M and v is positive on M^+.

In addition, define

$$\mathfrak{E}^0_\tau = \{r \in \mathfrak{E}_\tau | (\log v)\sigma \text{ is integrable over } M < r >\}.$$

$\mathfrak{E}^0_\tau$ is independent of the choice of Ψ and $\mathbb{R}^+ - \mathfrak{E}^0_\tau$ has measure zero. See p. 42 of Stoll [12].

Then, for $0 < s < r$ in $\mathfrak{E}^0_\tau$ and s in $\mathfrak{E}^0_\tau$, the Ricci function of τ is given by

$$Ric_\tau(r, s) = \int_{M<r>} (\log v)\sigma - \int_{M<s>} (\log v)\sigma + Ric(r, s, \Psi) \tag{1.5}$$

and is independent of the choice of Ψ.

THEOREM 1.1 *Let τ be a parabolic exhaustion of M. Let $\Theta \not\equiv 0$ be a holomorphic form of degree m. Let β be the zero-divisor of Θ. On M^+, define a function $\mathbb{Z} \geq 0$ such that $\mathbb{Z}^2$ is of class C^∞ on M^+ and $\mathbb{Z}^2\boldsymbol{v}^m = i_m\Theta \wedge \overline{\Theta}$. On $M - M^+$, set $\mathbb{Z} = 1$. For r in $\mathfrak{E}_\tau$, then r is in $\mathfrak{E}^0_\tau$ if and only if $(\log \mathbb{Z})\sigma$ is integrable over $M < r >$. For s in $\mathfrak{E}^0_\tau$ and r in $\mathfrak{E}^0_\tau$, then*

$$\mathrm{Ric}_\tau(r, s) + \int_{M<r>} (\log \mathbb{Z})\sigma - \int_{M<s>} (\log \mathbb{Z})\sigma = N_\beta(r, s). \tag{1.6}$$

PROOF: See Lemma 15.3 of Stoll [10].

1.2 Jacobian sections for complex manifolds

Let M and N be connected complex manifolds of dimension m and n, respectively. Let $f : M \to N$ be a holomorphic map. Also, suppose that N is compact. Let $K_{N,f}$ be the pullback of the canonical bundle K_N and $K^*_{N,f}$ the dual of the pullback. Then, the Jacobian bundle of f is $K(f) = K_M \otimes K^*_{N,f}$. A Jacobian section F is a global holomorphic section of $K(f)$. F is called effective if $Z(F)$ is thin. See Bardis [2], p. 5.

Now, for U an open subset of N, consider $\Omega^n_N(U)$, the vector space of holomorphic forms of degree n on U. Let $\tilde{U} = f^{-1}(U)$ be nonempty. F acts as a homomorphism mapping $\Omega^n_N(U)$ to $\Omega^m_M(\tilde{U})$. Suppose $(\cdot, \cdot)$ is the inner product between $K(f)$ and $K_{N,f}$ yielding values in K_M. Suppose ψ is in $\Omega^n_N(U)$. ψ can be pulled back to a holomorphic section ψ_f of $K_{N,f}$ over $\tilde{U}$. Then, in $\Omega^m_M(U)$,

$$F[\psi] = (F, \psi_f).$$

Here, if $Z[\psi]$ is empty, then $Z(F[\psi]) = Z(F) \cap \tilde{U}$.

Suppose that $\alpha : U_\alpha \to U'_\alpha$ and $\beta : U_\beta \to U'_\beta$ are smooth charts defined on M and N, respectively, such that U_β is contained in U and $f(U_\alpha)$ is contained in U_β. Then, on U_β, $\psi = \psi_\beta d\beta$ and on U_α, $F = F_{\alpha\beta} d\alpha \otimes d^*\beta_f$ for some holomorphic functions ψ_β and $F_{\alpha\beta}$. Then, on U_α, define

$$F[\psi] \quad = \quad F_{\alpha\beta}(\psi_\beta \circ f) d\alpha.$$

The operator $F : \Omega^n_N \to \Omega^m_M$ is called a holomorphic volume homomorphism if it meets the following conditions:
(E1) For ψ in Ω^n_N and χ in Ω^n_N, $F[\psi + \chi] = F[\psi] + F[\chi]$.
(E2) For ψ in Ω^n_N and h in Ω^0_N, $F[h\psi] = (h \circ f)F[\psi]$.
(E3) For V an open subset of U, $\tilde{V} = f^{-1}(V)$ nonempty, and ψ in Ω^n_N, $F[\psi]|\tilde{V} = F[\psi|V]$.
Such a holomorphic volume homomorphism F uniquely defines a Jacobian section $\tilde{F}$ with $\tilde{F}[\psi] = F[\psi]$.

Furthermore, let $A^p_N(U)$ (respectively $A^{p,q}_N(U)$) denote the C^∞ forms of degree p (respectively bidegree (p, q)) on U. Then, the action of F can be extended to a homomorphism mapping $\Omega^{2n}_N(U)$ to $\Omega^{2m}_M(\tilde{U})$. For ϕ in $\Omega^n_N(U)$, χ in $\Omega^n_N(U)$, and i_ρ given as in the beginning notation, define

$$F[i_n \phi \wedge \overline{\chi}] \quad = i_m F[\phi] \wedge \overline{F[\chi]}.$$

See p. 327 of Stoll [11] for further explanation.

If $f : M \to N$ is a holomorphic map and τ is a parabolic exhaustion of M, then the Jacobian section F of f is said to be dominated by τ if and only if for each $r > 0$ there exists a constant $C \geq 0$ such that the following property (D) holds:

(D) If U is an open subset of N, $\tilde{U} = f^{-1}(U)$, and $\tilde{U} \cap M^+(r)$ is nonempty, then

$$n \left(\frac{F[\psi]^n}{\boldsymbol{v}^m} \right)^{\frac{1}{n}} \boldsymbol{v}^m \quad \leq \quad C f^*(\psi) \wedge \boldsymbol{v}^{m-1} \tag{1.7}$$

is valid on $\tilde{U} \cap M^+(r)$ for all continuous, nonnegative forms ψ in $A^{1,1}_M(U)$. See p. 328 of Stoll [11]. Since τ is parabolic, $M^+(r)$ is nonempty for all $r > 0$.

Denote $Y(r)$ as the infimum of all constants C such that 1.7 holds. Then, in fact, $Y(r) \geq 0$ and on $\tilde{U} \cap M^+(r)$, $Y(r)$ satisfies

$$n \left(\frac{F[\psi]^n}{\boldsymbol{v}^m} \right)^{\frac{1}{n}} \boldsymbol{v}^m \quad \leq \quad Y(r) f^*(\psi) \wedge \boldsymbol{v}^{m-1}. \tag{1.8}$$

Here, the function $Y = Y_F$ is called the dominator of F. Note that Y is an increasing function of r. For more details about Jacobian sections, see pp. 327-328 of Stoll [11] and pp. 109-118 of Stoll [10].

Now, the concept of a Jacobian section needs to be extended to meromorphic maps $f : M \to N$. Suppose I_f is the indeterminacy of f. Then, the dimension of I_f is at most $m - 2$. The closed graph $M_f = \overline{\{(z, f(z)) | z \in M - I_f\}}$ is an irreducible analytic subset of $M \times N$ of dimension m. Also, the projections $\tilde{\pi} : M_f \to M$ and $\tilde{f} : M_f \to N$ are holomorphic such that $\tilde{\pi}$ is surjective and proper. Define $\tilde{\pi}^{-1}(I_f) = \tilde{I}_f$. Then, the mapping $\hat{\pi} : M_f - \tilde{I}_f \to M - I_f$ is biholomorphic and $f_0 = f|(M - I_f)$ is holomorphic. The Jacobian section F for f_0 is called Jacobian for f if and only if there exists a Jacobian section $\tilde{F}$ of M_f for $\tilde{f}$ such that the following property (P) holds:

(P) Let U be a nonempty, open subset of N such that $\hat{U} = \tilde{f}^{-1}(U)$ is nonempty. Then, $\hat{U} - \tilde{I}_f$ is nonempty. Define $\tilde{U} = \hat{\pi}(\hat{U} - \tilde{I}_f) = f_0^{-1}(U)$. For χ in $\Omega^n_N(U)$,

$$\hat{\pi}^*(F[\chi]) = \tilde{F}[\chi]|(\hat{U} - \tilde{I}_f). \tag{1.9}$$

Suppose $q = m - n \geq 0$ and take $\phi \not\equiv 0$ in $\Omega^q_M(M)$. Then, $0 \not\equiv \tilde{\phi} = \tilde{\pi}^*(\phi)$ in $\Omega^q_{M_f}(M_f)$ and $0 \not\equiv \phi_0 = \phi|(M - I_f)$ in $\Omega^q_M(M - I_f)$. In turn, the Jacobian sections F_{ϕ_0} of $M - I_f$ for f_0 and $F_{\tilde{\phi}}$ of M_f for $\tilde{f}$ can be defined. It can then be shown that property (P) holds with $\tilde{F}_{\phi_0} = F_{\tilde{\phi}}$ and that F_{ϕ_0} is a Jacobian section on M denoted by F_ϕ. See Chapter 1, Section 1 and, in particular, pp. 7-8 of Bardis [2] for more details.

1.3 Preparations

To begin with, the assumptions behind the Carlson-Griffiths method need to be introduced. This then leads to the crucial result found on line 1.12 used in the proof of the Lemma of the Logarithmic Derivative. To see the details needed in this result, see Chapter 1, Section 2 of Bardis [2].

ASSUMPTIONS:

[1] N is a compact, connected, projective algebraic manifold of dimension n.
[2] κ_0 is a hermitian metric on K^*_N with $c(K^*_N, \kappa_0) > 0$.
[3] ψ is a meromorphic (n,0) form on N without zeroes.

[4] $\mu^\infty_\psi \geq 0$ has strictly normal crossings. Let $q - 1$ be the number of branches of $\text{supp}\mu^\infty_\psi$.

[5] (M, τ) is a parabolic manifold of dimension $m > 0$.
[6] f: $M \to N$ is a meromorphic map with indeterminacy I_f.
[7] F is an effective Jacobian section for f dominated by τ with dominator Y_F.

Denote $f_0 = f|(M - I_f)$ where I_f is the indeterminacy of f. Then, define $\Delta \geq 0$ on M^+ by

$$\Delta^2 \boldsymbol{v}^m = i_m F[\psi] \wedge F[\overline{\psi}]. \tag{1.10}$$

Also, for almost all $r > 0$, define

$$D(r) = \int_{M<r>} (\log^+ \Delta)\sigma. \tag{1.11}$$

Now, the following important result can be stated.

THEOREM 1.2 *Suppose assumptions* [1]-[7] *are valid and* $\epsilon > 0$, $0 < s < r$. *Then,*

$$D(r) \mathrel{\underset{\cdot\cdot}{\leq}} (1 + \epsilon)\left(\frac{n}{2} + 2q\right)\varsigma \log T_f(r, s, K^*_N, \kappa_0) + (1 + \epsilon)n\varsigma(\log^+ Y_F(r) + 6\log^+ r). \tag{1.12}$$

PROOF: See Theorem 1.2.$\tilde{9}$ of Bardis [2]. Here, $\underset{\cdot\cdot}{\leq}$ means that the inequality is valid except for a set of finite measure in $\mathbb{R}^+$.

As done on pp. 43 and 152 of Stoll [12], a differential operator needs to be introduced on the connected, complex manifold M of dimension m for $m > 1$. To begin with, let $\mathbb{B}$ be a holomorphic $(m-1,0)$ form defined on M. Then, $\mathbb{B}$ defines a differential operator $'$ on every chart $\mathfrak{z} = (z_1, ..., z_m) : U_{\mathfrak{z}} \to U'_{\mathfrak{z}}$ of M. Define $\zeta = dz_1 \wedge ... \wedge dz_m$. Then, if $\mathfrak{v}$ is any holomorphic vector function on $U_{\mathfrak{z}}$, define $\mathfrak{v}'$ by

$$\mathfrak{v}'\zeta \quad = \quad d\mathfrak{v} \wedge \mathbb{B}.$$

Here, $\mathfrak{v}'$ is called the <u>$\mathbb{B}$-derivative</u> of $\mathfrak{v}$. On the open neighborhood U there exist unique holomorphic functions B_j satisfying

$$\mathbb{B} \quad = \quad \sum_{j=1}^{m} (-1)^{j-1} B_j dz_1 \wedge ... \wedge dz_{j-1} \wedge dz_{j+1} \wedge ... \wedge dz_m.$$

In particular, $\mathfrak{v}'$ can be explicitly expressed by

$$\mathfrak{v}' \quad = \quad \sum_{j=1}^{m} B_j \frac{\partial \mathfrak{v}}{\partial z_j}.$$

Notice that the operator $'$ can be iterated.

Next, assume that there exists a holomorphic form Θ of bidegree $(m,0)$ on M, where $\Theta \not\equiv 0$. In other words, there exists a nontrivial section Θ of the canonical bundle on M. This condition is an important one and the form Θ will remain fixed throughout this paper. Similarly, the form $\mathbb{B}$ will be fixed throughout this paper and will satisfy certain conditions necessary for the derivation of the defect relation.

Examples of manifolds on which such a nontrivial section Θ exists include all parabolic Stein manifolds. In particular, for noncompact, parabolic Riemann surfaces M_i, $1 \leq i \leq m$, this condition is met on $M_1 \times ... \times M_m$.

Next, a global meromorphic differential operator D is defined by

$$DH \cdot \Theta = dH \wedge \mathbb{B} \not\equiv 0, \tag{1.13}$$

where H is a meromorphic vector function. D is used as a differential operator mapping functions to functions and can be iterated such that $D^2(H) = D(DH)$ and so on. It is worthy of note that in the case of Bardis [2], $\Theta = d\pi_1 \wedge ... \wedge d\pi_m$ where $\pi : M \to \mathbb{C}^m$ is a covering space. Given a covering parabolic manifold, this is a natural choice for Θ.

Chapter 2

2.1 Lemma of the logarithmic derivative on parabolic manifolds

Theorem 1.2 will be applied to prove the Lemma of the Logarithmic Derivative for Functions. Before this can be done, some preparatory work must be completed.

Make the following assumptions:

ASSUMPTIONS A

[A1] Let (M, τ) be a parabolic manifold of dimension $m > 0$.

[A2] Let V be a complex vector space of dimension $n + 1 > 1$.

[A3] Let f: $M \to \mathbb{P}(V)$ be a meromorphic map.

[A4] Let $\mathbb{B}$ be a holomorphic form of bidegree $(m-1, 0)$ on M.

[A5] Let f be general for $\mathbb{B}$. In other words, if $\mathfrak{v} : U \to V$ is a reduced representation of f taken on a chart $\mathfrak{z} : U \to U'$, then $\mathfrak{v}_{\underline{n}} : U \to \tilde{G}_n(V)$ which is given by $\mathfrak{v}_{\underline{n}} = \mathfrak{v} \wedge \mathfrak{v}' \wedge ... \wedge \mathfrak{v}^{(n)}$ is such that $\mathfrak{v}_{\underline{n}} \not\equiv 0$.

[A6] Let $\mathbb{B}$ be majorized by τ with majorant $Y_{\mathbb{B}}$. In other words, on $M[r]$ for each $r > 0$, there exists a constant $c \geq 0$ such that

$$mi_{m-1}\mathbb{B} \wedge \overline{\mathbb{B}} \quad \leq \quad c\boldsymbol{v}^{m-1}.$$

Let $Y_{\mathbb{B}}(r)$ be the infimum of these constants. Notice that $Y_{\mathbb{B}}(r)$ is increasing. Then,

$$mi_{m-1}\mathbb{B} \wedge \overline{\mathbb{B}} \quad \leq \quad Y_{\mathbb{B}}(r)\boldsymbol{v}^{m-1}. \tag{2.1}$$

See p. 47 of Stoll [12].

[A7] Let $\Theta \not\equiv 0$ be a holomorphic form of bidegree $(m, 0)$ on M. [A8] Let β be the zero divisor of Θ. Let κ be the hermitian metric along the fibers of $\bigwedge_p \mathfrak{T}^*(M)$

defined by $\boldsymbol{\upsilon} > 0$ over M^+ for each $0 < p \leq m$. Take φ in $\underset{p}{\wedge} \mathfrak{T}^*(M)_x$ and χ in $\underset{p}{\wedge} \mathfrak{T}^*(M)_x$. Then, the hermitian product $(\varphi|\chi)_\kappa$ is defined by

$$(\varphi|\chi)_\kappa \boldsymbol{\upsilon}^m = \binom{m}{p} i_p \varphi \wedge \overline{\chi} \wedge \boldsymbol{\upsilon}^{m-p}. \tag{2.2}$$

Consequently,

$$\| \Theta \|_\kappa^2 \, \boldsymbol{\upsilon}^m = i_m \Theta \wedge \overline{\Theta}. \tag{2.3}$$

Note that the definition of $\mathbf{Z}$ in Theorem 1.1 implies that $\mathbf{Z} = \| \Theta \|_\kappa$ on M^+.

Now, several results concerning Jacobian sections need to be introduced.

CONSTRUCTION 2.0 Suppose that [A1], [A4], and [A6] are valid. Also, suppose that H is a meromorphic function on M such that

$$dH \wedge \mathbb{B} \not\equiv 0. \tag{2.4}$$

H will be viewed as a meromorphic function and a meromorphic map

$$H : M \to \mathbb{P}_1.$$

Under both interpretations, the set of indeterminacy I_H will be the same such that the dimension of I_H is at most $m - 2$.

Viewing H as a map, the closed graph of H

$$M_H = \overline{\{(z, H(z)) \in M \times \mathbb{P}_1 | z \in M - I_H\}}$$

is an m-dimensional, irreducible analytic set in $M \times \mathbb{P}_1$. Let $\iota : M_H \to M \times \mathbb{P}_1$ be the inclusion map. Let $\ddot{\pi} : M \times \mathbb{P}_1 \to M$ and $\ddot{\eta} : M \times \mathbb{P}_1 \to \mathbb{P}_1$ be the projections of $M \times \mathbb{P}_1$. Then, define the holomorphic projections of M_H by

$$\tilde{\pi} = \ddot{\pi} \circ \iota : M_H \to M \qquad \text{and} \qquad \tilde{\eta} = \ddot{\eta} \circ \iota : M_H \to \mathbb{P}_1.$$

Here, $\tilde{\pi}$ is both proper and surjective. Also, $\tilde{I}_H = \tilde{\pi}^{-1}(I_H)$ is a thin, analytic subset of M_H. The restriction map

$$\hat{\pi} = \tilde{\pi} : M_H - \tilde{I}_H \to M - I_H$$

is biholomorphic such that for all z in $M - I_H$,

$$\hat{\pi}^{-1}(z) = (z, H(z)).$$

Next, the maps

$$H_0 = H : M - \tilde{I}_H \to \mathbb{P}_1 \qquad \text{and} \qquad \eta_0 = \tilde{\eta} : M_H - \tilde{I}_H \to \mathbb{P}_1$$

are holomorphic. For z in $M - I_H$,

$$(\eta_0 \circ \hat{\pi}^{-1})(z) = \eta_0(z, H(z)) = H(z) = H_0(z).$$

Therefore,

$$\eta_0 \circ \hat{\pi}^{-1} = H_0 \qquad \text{or} \qquad \eta_0 = H_0 \circ \hat{\pi}.$$

A Jacobian section F is defined for $H_0 : M - I_H \to \mathbb{P}_1$ in the following way. Let U be an open subset of $\mathbb{P}_1$ on which is defined a holomorphic form ϕ of bidegree (1,0). Also, suppose that $\emptyset \neq \tilde{U} = H_0^{-1}(U)$ is an open subset of $M - I_H$. Then,

$$F[\phi] \quad = \quad H_0^*(\phi) \wedge \mathbb{B}.$$

Here, $F[\phi]$ is in $\Omega^m(H_0^{-1}(U)) = \Omega^m(\tilde{U})$. Furthermore, F satisfies conditions (E1) through (E3) of Section 1.2 and thus can be viewed as a Jacobian section. Additionally, F is effective since on $\tilde{U}$,

$$\begin{aligned} F[dz] &= H_0^*(dz) \wedge \mathbb{B} \\ &= d(H_0|\tilde{U}) \wedge \mathbb{B} \\ &= d(H|\tilde{U}) \wedge \mathbb{B} \not\equiv 0 \end{aligned}$$

due to 2.4.

Next, set $\tilde{\mathbb{B}} = \tilde{\pi}(\mathbb{B})$. Let $\hat{U} = \tilde{\eta}^{-1}(U) = (U \times \mathbb{P}_1) \cap M_H$ be a nonempty subset of M_H. Then, $\hat{U} - \tilde{I}_H$ is nonempty. For ϕ in $\Omega^1_{\mathbb{P}_1}(U)$, define the Jacobian section $\tilde{F}$ for $\tilde{\eta} : M_H \to \mathbb{P}_1$ by

$$\tilde{F}[\phi] \quad = \quad \tilde{\eta}^*(\phi) \wedge \tilde{\mathbb{B}}.$$

Again, it can be verified that $\tilde{F}$ satisfies (E1) through (E3) and is indeed a Jacobian section of M_H such that $\tilde{F}[\phi]$ is in $\Omega^m(\hat{U})$. Now, 1.9 must be checked. Take χ in $\Omega^1_{\mathbb{P}_1}(U)$. It must be shown that

$$\hat{\pi}^*(F[\chi]) \quad = \quad \tilde{F}[\chi]|(\hat{U} - \tilde{I}_H).$$

It must first be noted that $\hat{\pi}^*(\tilde{U}) = \hat{U} - \tilde{I}_H$. Then,

$$\begin{aligned} \hat{\pi}^*(F[\chi]) &= \hat{\pi}^*(H_0^*(\chi) \wedge \mathbb{B}) \\ &= (H_0 \circ \hat{\pi})^*(\chi) \wedge \hat{\pi}^*(\mathbb{B}) \\ &= \eta_0^*(\chi) \wedge \tilde{\mathbb{B}} \\ &= \tilde{\eta}^*(\chi) \wedge \tilde{\mathbb{B}}|(\hat{U} - \tilde{I}_H) \\ &= \tilde{F}[\chi] \wedge \tilde{\mathbb{B}}|(\hat{U} - \tilde{I}_H). \end{aligned}$$

Now, consider the following lemma:

LEMMA 2.1 *Suppose assumptions* [A1], [A4], *and* [A6] *are valid. Let H be a meromorphic function on M with $dH \wedge \mathbb{B} \not\equiv 0$. Let F be the Jacobian section for H constructed in 2.0. If $r > 0$, then*

$$Y_F(r) \leq Y_{\mathbb{B}}(r). \tag{2.5}$$

PROOF: Let U be an open subset of $\mathbb{P}_1$. Also, let $\tilde{U} = H_0^{-1}(U)$ which is contained in $M - I_H$. Assume that $\tilde{U} \cap M^+(r)$ is nonempty. Here, $n = 1$. Let χ be a nonnegative, continuous form of bidegree (1,1) defined on U. Then, by 1.8,

$$F[\chi] \quad \leq \quad Y_F(r) H_0^*(\chi) \wedge \boldsymbol{v}^{m-1} \tag{2.6}$$

on $\tilde{U} \cap M^+(r)$.

Here, $Y_F(r)$ is the smallest nonnegative constant for which this inequality holds for all permissible choices of U and χ.

Now, let $P = H^{-1}(\infty)$. On $U - \{\infty\}$, for $\tilde{\chi}$ a nonnegative, continuous function,

$$\chi = \tilde{\chi}\frac{i}{2\pi}dz \wedge d\bar{z}.$$

Then, on $\tilde{U} \cap (M^+(r) - P)$,

$$\begin{aligned} F[\chi] &= i_m(\tilde{\chi} \circ H_0)F[dz] \wedge \overline{F[dz]} \\ &= i_m(\tilde{\chi} \circ H_0)dH_0 \wedge \mathbb{B} \wedge \overline{dH_0} \wedge \overline{\mathbb{B}} \\ &= (\tilde{\chi} \circ H_0)\frac{i}{2\pi}dH_0 \wedge \overline{dH_0} \wedge m i_{m-1}\mathbb{B} \wedge \overline{\mathbb{B}} \\ &= H_0^*(\chi) \wedge m i_{m-1}\mathbb{B} \wedge \overline{\mathbb{B}}. \end{aligned}$$

By continuity, this equality extends to $\tilde{U} \cap M^+(r)$:

$$F[\chi] = H_0^*(\chi) \wedge m i_{m-1}\mathbb{B} \wedge \overline{\mathbb{B}}. \tag{2.7}$$

From p. 141 of Stoll [10],

$$m i_{m-1}\mathbb{B} \wedge \overline{\mathbb{B}} \leq Y_{\mathbb{B}}(r)\boldsymbol{v}^{m-1}.$$

Combining this fact and 2.7, on $\tilde{U} \cap M^+(r)$,

$$F[\chi] \leq Y_{\mathbb{B}}(r)H_0^*(\chi) \wedge \boldsymbol{v}^{m-1}. \tag{2.8}$$

Hence, by 2.6 and 2.8 and for $r > 0$,

$$Y_F(r) \leq Y_{\mathbb{B}}(r).$$

q.e.d.

THEOREM 2.2 [LEMMA OF THE LOGARITHMIC DERIVATIVE FOR FUNCTIONS] *Suppose assumptions* [A1],[A4],[A6],[A7],[A8] *are valid. Let $H \not\equiv 0$ be a meromorphic function on M. Take $\epsilon > 0$ and $0 < s < r$. Then,*

$$\int_{M\langle r\rangle} \log^+\left(\left|\frac{DH}{H}\right|\mathbf{Z}\right)\sigma \underset{\cdot\cdot}{\leq} (1+\epsilon)\varsigma[7\log T_H(r,s) + \log^+ Y_{\mathbb{B}}(r) + 6\log^+ r]. \tag{2.9}$$

PROOF: First of all, notice that 2.9 will hold trivially if $dH \wedge \mathbb{B} \equiv 0$ since $dH \wedge \mathbb{B} = DH \cdot \Theta$ by definition. Thus, it can be assumed that $dH \wedge \mathbb{B} \not\equiv 0$.

Theorem 1.2 will be applied via the following translation table:

T1.2	N	n	κ_0	ψ	μ_ψ^∞	q	(M,τ)	m	f	I_f	F	Y_F
Here	$\mathbb{P}_1$	1	κ_1^2	$\frac{dz}{z}$	μ_ψ^∞	3	(M,τ)	m	H	I_H	F	Y_F

T1.2	$c(K_N^*,\kappa_0)$	Δ	$D(r)$	$T_f(r,s,K_N^*,\kappa_0)$
Here	2Ω	$\lvert\frac{DH}{H}\rvert\mathbf{Z}$	$D(r)$	$2T_H(r,s)$

Here κ_1 is the standard hermitian metric along the fibers of $\mathcal{O}(1)$ and $\Omega = c(\mathcal{O}(1), \kappa_1)$. Thus, $K_N = \mathcal{O}(-2)$ while $K_N^* = \mathcal{O}(2)$ and $\kappa_0 = \kappa_1^2$ implies that $c(K_N^*, \kappa_0) = c(\mathcal{O}(2), \kappa_1^2) = 2c(\mathcal{O}(1), \kappa_1) = 2\Omega > 0$. In addition, $\psi(z) = \frac{dz}{z}$ is a meromorphic form of bidegree $(1,0)$ without zeroes and

$$\mu_\psi^\infty(z) = \begin{cases} 0 & 0 \neq z \neq \infty \\ 1 & z = 0 \\ 1 & z = \infty. \end{cases}$$

Therefore, $q = 3$. H is a meromorphic function and thus a meromorphic map $H\colon M \to \mathbb{P}_1$. Assume that $H \not\equiv 0$ and $dH \wedge \mathbb{B} \not\equiv 0$. Then, by 1.10 and 1.13,

$$\begin{aligned} \Delta^2 \boldsymbol{v}^m &= i_m F\left[\frac{dz}{z}\right] \wedge \overline{F\left[\frac{dz}{z}\right]} \\ &= i_m \frac{dH}{H} \wedge \mathbb{B} \wedge \overline{\left(\frac{dH}{H}\right)} \wedge \overline{\mathbb{B}} \\ &= i_m \left|\frac{DH}{H}\right|^2 \Theta \wedge \overline{\Theta} \\ &= \left|\frac{DH}{H}\right|^2 \parallel \Theta \parallel_\kappa^2 \boldsymbol{v}^m. \end{aligned}$$

Hence,

$$\Delta = \left|\frac{DH}{H}\right| \parallel \Theta \parallel_\kappa. \tag{2.10}$$

By 1.11 and 2.10,

$$D(r) = \int_{M\langle r\rangle} \left(\log^+ \left|\frac{DH}{H}\right| \parallel \Theta \parallel_\kappa\right) \sigma.$$

Then, by applying 1.12 via the above translation table and using 2.5,

$$\begin{aligned} D(r) &\underset{\because}{\leq} (1+\epsilon)(\frac{1}{2} + 6)\varsigma \log(2T_H(r,s)) \\ &\quad + (1+\epsilon)\varsigma[\log^+ Y_{\mathbb{B}}(r) + 6\log^+ r] \\ &\underset{\because}{\leq} (1+\epsilon)\varsigma[7\log T_H(r,s) + \log^+ Y_{\mathbb{B}}(r) + 6\log^+ r]. \end{aligned}$$

Hence,

$$\int_{M\langle r\rangle} \left(\log^+ \left|\frac{DH}{H}\right| \mathbf{Z}\right) \sigma \underset{\because}{\leq} (1+\epsilon)\varsigma[7\log T_H(r,s) + \log^+ Y_{\mathbb{B}}(r) + 6\log^+ r].$$

q.e.d.

For all r in $\mathfrak{C}_\tau^0$, define

$$E(r) = \int_{M\langle r\rangle} (\log^+ \mathbf{Z})\sigma. \tag{2.11}$$

For r chosen in $\mathfrak{E}^0_\tau$, then $\int_{M<r>}(\log \mathbf{Z})\sigma$ exists due to 1.6. Hence, $\int_{M<r>} |\log \mathbf{Z}|\sigma$ exists. Since $0 \le \log^+ \mathbf{Z} \le |\log \mathbf{Z}|$, then $\int_{M<r>}(\log^+ \mathbf{Z})\sigma$ exists. Thus, $E(r)$ is well-defined.

Now, several properties of the $\log^+$ function need to be considered.

LEMMA 2.3 *If $x > 0$ and $y > 0$, then*

$$
\begin{aligned}
1.\quad \log \frac{\sqrt{1+x^2}}{1+y^2} &\le \log^+ \frac{x}{y}; && (2.12)\\
2.\qquad \log^+(xy) &\le \log^+ x + \log^+ y; && (2.13)\\
3.\ \log^+\left(\sum_{i=1}^n x_i\right) &\le \sum_{i=1}^n \log^+ x_i + \log n. && (2.14)
\end{aligned}
$$

PROOF: 1. It is enough to show that $\log \frac{1+x^2}{1+y^2} \le \log^+ \frac{x^2}{y^2}$. Two cases need to be considered. To begin with, if $|x| \le |y|$, then $x^2 \le y^2$ and $\frac{x^2}{y^2} \le 1$. This implies that $\log^+ \frac{x^2}{y^2} = 0$ while $\log \frac{1+x^2}{1+y^2} \le \log 1 = 0$. On the other hand, if $|x| > |y|$, then $x^2 > y^2$ and $\frac{x^2}{y^2} > 1$. In this case, $\frac{1+x^2}{1+y^2} \le \frac{x^2}{y^2}$ which implies that $\log \frac{1+x^2}{1+y^2} \le \log \frac{x^2}{y^2} = \log^+ \frac{x^2}{y^2}$.

2. Consider the following cases:
CASE 1: $x > 1, y > 1 \Longrightarrow \log^+ xy = \log xy = \log x + \log y = \log^+ x + \log^+ y$.
CASE 2: $x \le 1, y \le 1 \Longrightarrow \log^+ xy = 0$ and $\log^+ x = 0 = \log^+ y$.
CASE 3: $x < 1, y > 1, \frac{1}{x} < y \Longrightarrow \log^+ xy = \log xy = \log x + \log y < \log y$ while $\log^+ x = 0$ and $\log^+ y = \log y$.
CASE 4: $x < 1, y > 1, \frac{1}{x} \ge y \Longrightarrow \log^+ xy = 0$ while $\log^+ x = 0$ and $\log^+ y = \log y > 0$.
By symmetry, all of the possible cases have been proven.

3. Let $x_i = \max\{x_1, ..., x_n\}$. Then,

$$
\begin{aligned}
\log^+\left(\sum_{k=1}^n x_k\right) &\le \log^+ n x_i\\
&\le \log^+ x_i + \log n\\
&\le \sum_{k=1}^n \log^+ x_k + \log n.
\end{aligned}
$$

q.e.d.

THEOREM 2.4 *Assume* [A1],[A4],[A6],[A7],[A8] *hold. Let $H \not\equiv 0$ be a meromorphic function on M. Take $\epsilon > 0$. Then, for $0 < s < r$,*

$$
\begin{aligned}
T_{DH}(r,s) \;\underset{\cdot\cdot}{\le}\; & 2T_H(r,s) + Ric_\tau(r,s) + E(r) + (1+\epsilon)\varsigma[7\log T_H(r,s)\\
& + \log^+ Y_{\mathbb{B}}(r) + 7\log^+ r]. \qquad (2.15)
\end{aligned}
$$

PROOF: To begin with, notice that if $DH \cdot \Theta = dH \wedge \mathbb{B} \equiv 0$, then 2.15 is valid. Therefore, assume that $dH \wedge \mathbb{B} \not\equiv 0$.

Let $\mathfrak{U} = \{U_\lambda\}_{\lambda\in\Lambda}$ be an open covering of M such that the following hold:

1. U_λ is nonempty, open, and connected for all λ in Λ.
2. $\mathfrak{u}_\lambda = (h_\lambda, g_\lambda)$ is a reduced representation of H such that $H|U_\lambda = \frac{g_\lambda}{h_\lambda}$ for all λ in Λ.
3. $\tilde{\mathfrak{u}}_\lambda$ is a holomorphic frame of the trivial bundle $\mathbb{C}^2_M = M \times \mathbb{C}^2$ such that $\tilde{\mathfrak{u}}_\lambda(x) = (x, \mathfrak{u}_\lambda(x))$ for all x in U_λ and λ in Λ.
4. There exists a nowhere zero holomorphic function $g_{\lambda\mu}$ such that for each (λ, μ) in $\Lambda[1]$, $\mathfrak{u}_\lambda = g_{\lambda\mu}\mathfrak{u}_\mu$ and $\tilde{\mathfrak{u}}_\lambda = g_{\lambda\mu}\tilde{\mathfrak{u}}_\mu$.
5. $\overset{\triangle}{\mathfrak{u}}_\lambda$ is a holomorphic frame of the hyperplane section bundle L_H for H on U_λ for all λ in Λ. Also, $\overset{\triangle}{\mathfrak{u}}_\mu = g_{\mu\lambda}\overset{\triangle}{\mathfrak{u}}_\lambda$ on $U_{\lambda\mu}$ for all (λ, μ) in $\Lambda[1]$.
6. The <u>canonical representation section</u> $F \not\equiv 0$ is a holomorphic section of $(M, \mathbb{C}^2_M \times L_H)$ uniquely defined by $F|U_\lambda = \tilde{\mathfrak{u}}_\lambda \otimes \overset{\triangle}{\mathfrak{u}}_\lambda$ for all λ in Λ.
7. ζ_λ is a holomorphic frame of the canonical bundle K_M over U_λ for all λ in Λ.
8. $\Theta_\lambda \not\equiv 0$ is holomorphic on U_λ such that $\Theta = \Theta_\lambda\zeta_\lambda$ on U_λ for all λ in Λ.

Then, $\mathfrak{u}_\lambda = (h_\lambda, g_\lambda) = (h_\mu g_{\lambda\mu}, g_\mu g_{\lambda\mu}) = g_{\lambda\mu}\mathfrak{u}_\mu$ which implies that

$$h_\lambda = h_\mu g_{\lambda\mu} \text{ and } g_\lambda = g_\mu g_{\lambda\mu}.$$

Applying D to these two equations,

$$Dh_\lambda = h_\mu Dg_{\lambda\mu} + g_{\lambda\mu}Dh_\mu \text{ and } Dg_\lambda = g_\mu Dg_{\lambda\mu} + g_{\lambda\mu}Dg_\mu.$$

Define

$$W_\lambda = \begin{vmatrix} h_\lambda & g_\lambda \\ Dh_\lambda & Dg_\lambda \end{vmatrix}.$$

Expanding this expression,

$$\begin{aligned} W_\lambda &= h_\lambda Dg_\lambda - g_\lambda Dh_\lambda \\ &= (h_\mu g_{\lambda\mu})(g_\mu Dg_{\lambda\mu} + g_{\lambda\mu}Dg_\mu) - (g_\mu g_{\lambda\mu})(h_\mu Dg_{\lambda\mu} + g_{\lambda\mu}Dh_\mu) \\ &= (g_{\lambda\mu})^2[h_\mu Dg_\mu - g_\mu Dh_\mu]. \end{aligned}$$

Hence, $W_\lambda = (g_{\lambda\mu})^2 \begin{vmatrix} h_\mu & g_\mu \\ Dh_\mu & Dg_\mu \end{vmatrix} = (g_{\lambda\mu})^2 W_\mu$.

Now,

$$DH = D\left(\frac{g_\lambda}{h_\lambda}\right) = \frac{h_\lambda Dg_\lambda - g_\lambda Dh_\lambda}{{h_\lambda}^2} = \frac{W_\lambda}{{h_\lambda}^2}.$$

Let $\mathfrak{w}_\lambda = ({h_\lambda}^2, W_\lambda)$ be a meromorphic representation of DH on U_λ. By the transition formulas for h_λ and W_λ,

$$\mathfrak{w}_\lambda = ({h_\lambda}^2, W_\lambda) = ({h_\mu}^2 {g_{\lambda\mu}}^2, W_\mu {g_{\lambda\mu}}^2) = {g_{\lambda\mu}}^2({h_\mu}^2, W_\mu) = {g_{\lambda\mu}}^2\mathfrak{w}_\mu.$$

Thus, $L_H^2 = L_{DH}$ where L_H has the transition functions $g_{\lambda\mu}$ while L_{DH} has the transition functions ${g_{\lambda\mu}}^2$. A meromorphic representation section $G \not\equiv 0$ in $\Gamma(M, \mathbb{C}_M^2 \otimes L_H^2)$ of DH is given by $G|U_\lambda = \tilde{\mathfrak{w}}_\lambda \otimes (\overset{\triangle}{\mathfrak{u}_\lambda})^2$.

By assumption, there exists a holomorphic function Θ_λ such that $\Theta = \Theta_\lambda \zeta_\lambda$ where ζ_λ is a holomorphic frame of K_M. Therefore,

$$\begin{aligned} dg_\lambda \wedge \mathbb{B} &= (Dg_\lambda)\Theta \\ &= (Dg_\lambda)\Theta_\lambda\zeta_\lambda; \\ dh_\lambda \wedge \mathbb{B} &= (Dh_\lambda)\Theta \\ &= (Dh_\lambda)\Theta_\lambda\zeta_\lambda. \end{aligned}$$

In addition,

$$\begin{aligned} h_\lambda dg_\lambda \wedge \mathbb{B} - g_\lambda dh_\lambda \wedge \mathbb{B} &= h_\lambda(Dg_\lambda)\Theta_\lambda\zeta_\lambda - g_\lambda(Dh_\lambda)\Theta_\lambda\zeta_\lambda \\ &= \Theta_\lambda\zeta_\lambda(h_\lambda Dg_\lambda - g_\lambda Dh_\lambda) \\ &= \Theta_\lambda\zeta_\lambda W_\lambda \\ &= W_\lambda\Theta_\lambda\zeta_\lambda. \end{aligned}$$

Thus, $W_\lambda\Theta_\lambda$ is a holomorphic function and $\Theta_\lambda\mathfrak{w}_\lambda = (\Theta_\lambda {h_\lambda}^2, \Theta_\lambda W_\lambda)$ is holomorphic. In turn, $\mu_G|U_\lambda = \mu_{\mathfrak{w}_\lambda} = \mu_{\Theta_\lambda\mathfrak{w}_\lambda} - \mu_{\Theta_\lambda} \geq -\mu_{\Theta_\lambda} = -\beta$. This then implies that

$$N_G(r,s) \geq -N_\beta(r,s). \tag{2.16}$$

Next, on $\mathbb{C}^2$ the standard Euclidean metric l defines a hermitian metric l on the fibers of $\mathbb{C}_M^2$. Also, choose a hermitian metric κ_2 on the fibers of L_H. Abbreviate $j = l \otimes \kappa_2$ and $k = l \otimes \kappa_2^2$.

By 2.16 and p. 140 of Stoll [12], the following relationships hold for $0 < s < r$:

$$\begin{aligned} T_H(r,s) &= \int_{M\langle r\rangle} \log \| F \|_j \,\sigma - \int_{M\langle s\rangle} \log \| F \|_j \,\sigma + T(r,s,L_H,\kappa_2); \\ 2T_H(r,s) &= \int_{M\langle r\rangle} \log \| F \|_j^2 \,\sigma - \int_{M\langle s\rangle} \log \| F \|_j^2 \,\sigma + \\ &\quad 2T(r,s,L_H,\kappa_2); \end{aligned} \tag{2.17}$$

$$\begin{aligned} T_{DH}(r,s) &= \int_{M\langle r\rangle} \log \| G \|_k \,\sigma - \int_{M\langle s\rangle} \log \| G \|_k \,\sigma + \\ &\quad T(r,s,L_H^2,\kappa_2^2) - N_G(r,s) \\ &\leq \int_{M\langle r\rangle} \log \| G \|_k \,\sigma - \int_{M\langle s\rangle} \log \| G \|_k \,\sigma + \\ &\quad 2T(r,s,L_H,\kappa_2) + N_\beta(r,s). \end{aligned} \tag{2.18}$$

Now, solving for $2T(r,s,L_H,\kappa_2)$ in 2.17 and then substituting in 2.18,

$$T_{DH}(r,s) \leq 2T_H(r,s) + \int_{M\langle r\rangle} \log \frac{\| G \|_k}{\| F \|_j^2}\sigma - \int_{M\langle s\rangle} \log \frac{\| G \|_k}{\| F \|_j^2}\sigma + N_\beta(r,s). \tag{2.19}$$

Furthermore, on U_λ,

$$\| G \|_k = \| \mathfrak{w}_\lambda \otimes (\overset{\triangle}{\mathfrak{u}_\lambda})^2 \| = \| \mathfrak{w}_\lambda \| \, \|\overset{\triangle}{\mathfrak{u}_\lambda}\|_{\kappa_2}^2;$$

$$\| F \|_j^2 = \| \mathfrak{u}_\lambda \otimes (\overset{\triangle}{\mathfrak{u}_\lambda})^2 \| = \| \mathfrak{u}_\lambda \|^2 \, \|\overset{\triangle}{\mathfrak{u}_\lambda}\|_{\kappa_2}^2.$$

Therefore,

$$\begin{aligned} \frac{\| G \|_k}{(\| F \|_j)^2} &= \frac{\| \mathfrak{w}_\lambda \|}{\| \mathfrak{u}_\lambda \|^2} \\ &= \frac{\sqrt{|h_\lambda^2|^2 + |W_\lambda|^2}}{|h_\lambda|^2 + |g_\lambda|^2} \\ &= \frac{\sqrt{1+|DH|^2}}{1+|H|^2}. \end{aligned}$$

Then, by 2.12,

$$\log \frac{\sqrt{1+|DH|^2}}{1+|H|^2} \leq \log^+ \frac{|DH|}{|H|}.$$

Applying this to 2.19,

$$\begin{aligned} T_{DH}(r,s) &\leq 2T_H(r,s) + \int_{M<r>} \log^+ \frac{|DH|}{|H|} \sigma + N_\beta(r,s) \\ &= 2T_H(r,s) + \int_{M<r>} \log^+ \frac{|DH|}{|H|} \sigma + Ric_\tau(r,s) + \\ &\quad \int_{M<r>} \log \mathbf{Z}\sigma - \int_{M<s>} \log \mathbf{Z}\sigma \end{aligned}$$

by 1.6. Using 2.13, notice that

$$\begin{aligned} \log^+ \frac{|DH|}{|H|} + \log \mathbf{Z} &= \log^+ \frac{|DH|\mathbf{Z}}{|H|\mathbf{Z}} + \log \mathbf{Z} \\ &\leq \log^+ \left(\frac{|DH|}{|H|} \mathbf{Z} \right) + \log^+ \frac{1}{\mathbf{Z}} + \log \mathbf{Z} \\ &= \log^+ \left(\frac{|DH|}{|H|} \mathbf{Z} \right) + \log^+ \mathbf{Z}. \end{aligned}$$

Therefore, for some constant $c(s)$,

$$\begin{aligned} T_{DH}(r,s) &\leq 2T_H(r,s) + Ric_\tau(r,s) + \int_{M<r>} \log^+ \mathbf{Z} + \\ &\quad \int_{M<r>} \log^+ \left(\frac{|DH|}{|H|} \mathbf{Z} \right) \sigma + c(s). \end{aligned}$$

Finally, by 2.9,

$$T_{DH}(r,s) \leq 2T_H(r,s) + Ric_\tau(r,s) + E(r) + \int_{M<r>} \log^+ \left(\frac{|DH|}{|H|} \mathbf{Z} \right) \sigma + c(s)$$

$$\begin{aligned} &\overset{..}{\leq} 2T_H(r,s) + Ric_\tau(r,s) + E(r) + (1+\epsilon)\varsigma[7\log T_H(r,s) \\ &\quad + \log^+ Y_{\mathbb{B}}(r) + 6\log^+ r] + c(s) \\ &\overset{..}{\leq} 2T_H(r,s) + Ric_\tau(r,s) + E(r) + (1+\epsilon)\varsigma[7\log T_H(r,s) + \\ &\quad \log^+ Y_{\mathbb{B}}(r) + 7\log^+ r]. \end{aligned}$$

q.e.d.

Now, a lemma of the logarithmic derivative for higher order derivatives needs to be derived. To begin with, several preliminary results must be introduced.

LEMMA 2.5 *For x in $\mathbb{R}$ and A, B in $\mathbb{R}^+$,*

$$\log^+(Ax + B\log^+ x) \leq \log^+ x + \log^+(A+B). \tag{2.20}$$

PROOF: If $x \leq 1$, then $\log^+ x = 0$ and therefore

$$\log^+(Ax + B\log^+ x) = \log^+ Ax \leq \log^+ A \leq \log^+ x + \log^+(A+B).$$

If $x > 1$, then

$$\begin{aligned} \log^+(Ax + B\log^+ x) &= \log^+(Ax + B\log x) \\ &\leq \log^+((A+B)x) \\ &\leq \log^+ x + \log^+(A+B). \end{aligned}$$

q.e.d.

LEMMA 2.6 *Take $0 < \epsilon$ in $\mathbb{R}$. Suppose $F : \mathbb{R}^+ \to \mathbb{R}$ is an increasing, unbounded function. Take A, B in $\mathbb{R}^+$. Then, there exists $0 < r_0$ in $\mathbb{R}$ such that*

$$(1+\frac{\epsilon}{2})(A\log^+ F(r) + BF(r)) < (1+\epsilon)BF(r) \tag{2.21}$$

for all $r \geq r_0$.

PROOF: Select r_1 in $\mathbb{R}^+$ such that $F(r) \geq 1$ for all $r \geq r_1$. Then,

$$\lim_{r\to\infty} (1+\frac{\epsilon}{2})\frac{A\log^+ F(r) + BF(r)}{F(r)} = (1+\frac{\epsilon}{2})B.$$

Thus, there exists $r_0 > r_1$ in $\mathbb{R}$ such that

$$(1+\frac{\epsilon}{2})\frac{A\log^+ F(r) + BF(r)}{F(r)} < (1+\epsilon)B$$

for all $r > r_0$. q.e.d.

Next, another preliminary result needs to be considered.

LEMMA 2.7 *Suppose assumptions* [A1],[A4],[A6],[A7],[A8] *are valid. Let $H \not\equiv 0$ be a meromorphic function on M such that $dH \wedge \mathbb{B} \not\equiv 0$. Take $0 < \epsilon$ in $\mathbb{R}$ and*

$0 < s < r$. *Then,*

$$\begin{aligned} T_{D^2H}(r,s) &\lesssim 4T_H(r,s) + 3[Ric_\tau(r,s) + E(r)] + (1+\epsilon)\varsigma[3\log^+ Y_{\mathbb{B}}(r) + \\ &\quad 22\log^+ r] + 7(1+\epsilon)\varsigma[3\log^+ T_H(r,s) + \\ &\quad \log^+ Ric_\tau(r,s) + \log^+ E(r)]; \\ \log^+ T_{D^2H}(r,s) &\lesssim \log^+ T_H(r,s) + \log^+ Ric_\tau(r,s) + \log^+ E(r) + \\ &\quad \log^+\log^+ Y_{\mathbb{B}}(r) + 2\log^+\log^+ r. \end{aligned}$$

PROOF: 2.15 states

$$\begin{aligned} T_{DH}(r,s) &\lesssim 2T_H(r,s) + Ric_\tau(r,s) + E(r) + (1+\frac{\epsilon}{2})\varsigma[7\log^+ T_H(r,s) \\ &\quad + \log^+ Y_{\mathbb{B}}(r) + 7\log^+ r]. \end{aligned}$$

By 2.20, there exists a constant $c(s)$ such that

$$\begin{aligned} \log^+ T_{DH}(r,s) &\lesssim \log^+ T_H(r,s) + \log^+ Ric_\tau(r,s) + \log^+ E(r) + \\ &\quad \log^+\log^+ Y_{\mathbb{B}}(r) + \log^+\log^+ r + c(s) \\ &\lesssim \log^+ T_H(r,s) + \log^+ Ric_\tau(r,s) + \log^+ E(r) + \\ &\quad \log^+\log^+ Y_{\mathbb{B}}(r) + 2\log^+\log^+ r. \end{aligned}$$

Then, since $D^2H = D(DH)$,

$$\begin{aligned} T_{D^2H}(r,s) &\lesssim 2T_{DH}(r,s) + Ric_\tau(r,s) + E(r) + \\ &\quad (1+\frac{\epsilon}{2})\varsigma[7\log T_H(r,s) + \log^+ Y_{\mathbb{B}}(r) + 7\log^+ r] \\ &\lesssim 4T_H(r,s) + 3[Ric_\tau(r,s) + E(r)] + \\ &\quad (1+\frac{\epsilon}{2})\varsigma[14\log^+ T_H(r,s) + 2\log^+ Y_{\mathbb{B}}(r) + 14\log^+ r] + \\ &\quad (1+\frac{\epsilon}{2})\varsigma[7\log^+ T_H(r,s) + 7\log^+ Ric_\tau(r,s) + 7\log^+ E(r) + \\ &\quad 7\log^+\log^+ Y_{\mathbb{B}}(r) + 14\log^+\log^+ r + \log^+ Y_{\mathbb{B}}(r) + 7\log^+ r] \\ &\lesssim 4T_H(r,s) + 3[Ric_\tau(r,s) + E(r)] + \\ &\quad 7(1+\frac{\epsilon}{2})\varsigma[3\log^+ T_H(r,s) + \log^+ Ric_\tau(r,s) + \log^+ E(r)] + \\ &\quad (1+\frac{\epsilon}{2})\varsigma[3\log^+ Y_{\mathbb{B}}(r) + 7\log^+\log^+ Y_{\mathbb{B}}(r) + 21\log^+ r + \\ &\quad 14\log^+\log^+ r] \\ &\lesssim 4T_H(r,s) + 3[Ric_\tau(r,s) + E(r)] + (1+\epsilon)\varsigma[3\log^+ Y_{\mathbb{B}}(r) + \\ &\quad 22\log^+ r] + 7(1+\epsilon)\varsigma[3\log^+ T_H(r,s) + \log^+ Ric_\tau(r,s) + \\ &\quad \log^+ E(r)]. \end{aligned}$$

Also, the estimate for $\log^+ T_{D^2H}(r,s)$ can be obtained in a way similar to the one for $\log^+ T_{DH}(r,s)$. In fact, if $Y_{\mathbb{B}}$ is not bounded, then only $21\log^+ r$ is needed. Otherwise, it is increased to $22\log^+ r$ to bound the term $7\log^+\log^+ Y_{\mathbb{B}}(r)$. q.e.d.

This then leads to a more general estimate of $T_{D^nH}(r,s)$ for n in $\mathbb{N}$.

THEOREM 2.8 *Suppose assumptions* [A1],[A4],[A6],[A7],[A8] *are valid. Once again, let* $H \not\equiv 0$ *be a meromorphic function on* M *such that* $dH \wedge \mathbb{B} \not\equiv 0$. *Take* $\epsilon > 0$,

$0 < s < r$, *and* n *in* $\mathbb{N}$. *Then,*

$$\begin{aligned} T_{D^nH}(r,s) \;\underset{\because}{\leq}\; & 2^n T_H(r,s) + (2^n-1)[Ric_\tau(r,s)+E(r)] + \\ & 7(1+\epsilon)\varsigma[(2^n-1)\log^+ T_H(r,s) + (2^{n-1}-1)(\log^+ Ric_\tau(r,s) + \\ & \log^+ E(r))] + (1+\epsilon)\varsigma[(2^n-1)\log^+ Y_{\mathbb{B}}(r) + \\ & (15(2^{n-1}-1)+7)\log^+ r]. \end{aligned}$$

PROOF: If $D^nH \equiv 0$, then $T_{D^nH}(r,s) \equiv 0$ and the estimate follows trivially. Therefore, assume that $T_{D^nH}(r,s) \not\equiv 0$. Hence, $T_{D^nH}(r,s) \to \infty$ as $r \to \infty$. From 2.15,

$$\begin{aligned} T_{DH}(r,s) \;\underset{\because}{\leq}\; & 2T_H(r,s) + Ric_\tau(r,s) + E(r) + (1+\frac{\epsilon}{2})\varsigma[7\log T_H(r,s) + \\ & \log^+ Y_{\mathbb{B}}(r) + 7\log^+ r]. \end{aligned}$$

In a similar fashion,

$$\begin{aligned} T_{D^nH}(r,s) \;\underset{\because}{\leq}\; & 2T_{D^{n-1}H}(r,s) + Ric_\tau(r,s) + E(r) + \\ & (1+\frac{\epsilon}{2})\varsigma[7\log T_{D^{n-1}H}(r,s) + \log^+ Y_{\mathbb{B}}(r) + 7\log^+ r]. \end{aligned}$$

For $n = 1$ or $n = 2$, this theorem is proven earlier in this chapter. Thus, assume that $n \geq 3$. Then, consider the following inductive hypothesis:

$$\begin{aligned} T_{D^{n-1}H}(r,s) \;\underset{\because}{\leq}\; & 2^{n-1}T_H(r,s) + (2^{n-1}-1)[Ric_\tau(r,s)+E(r)] + \\ & 7(1+\frac{\epsilon}{2})\varsigma[(2^{n-1}-1)\log^+ T_H(r,s) + \\ & (2^{n-2}-1)(\log^+ Ric_\tau(r,s) + \log^+ E(r))] + \\ & (1+\epsilon)\varsigma[(2^{n-1}-1)\log^+ Y_{\mathbb{B}}(r) + (15(2^{n-2}-1)+7)\log^+ r]. \end{aligned}$$

Note that because $n \geq 3$, $2^{n-2}-1 \geq 2-1 = 1$. Hence,

$$\begin{aligned} \log^+ T_{D^{n-1}H}(r,s) \;\underset{\because}{\leq}\; & \log^+ T_H(r,s) + \log^+ Ric_\tau(r,s) + \log^+ E(r) + \\ & \log^+\log^+ Y_{\mathbb{B}}(r) + 2\log^+\log^+ r. \end{aligned}$$

Thus, it follows that

$$\begin{aligned} T_{D^nH}(r,s) \;\underset{\because}{\leq}\; & 2^nT_H(r,s) + (2^n-1)[Ric_\tau(r,s)+E(r)] + \\ & 7(1+\frac{\epsilon}{2})\varsigma[(2^n-2)\log^+ T_H(r,s) + (2^{n-1}-2)(\log^+ Ric_\tau(r,s) + \\ & \log^+ E(r))] + (1+\frac{\epsilon}{2})\varsigma[(2^n-2)\log^+ Y_{\mathbb{B}}(r) + \\ & (15(2^{n-1}-2)+14)\log^+ r] + (1+\frac{\epsilon}{2})\varsigma[\log^+ Y_{\mathbb{B}}(r) + 7\log^+ r] + \\ & 7(1+\frac{\epsilon}{2})\varsigma[\log^+ T_H(r,s) + \log^+ Ric_\tau(r,s) + \log^+ E(r) + \\ & \log^+\log^+ Y_{\mathbb{B}}(r) + 2\log^+\log^+ r] \\ \underset{\because}{\leq}\; & 2^nT_H(r,s) + (2^n-1)[Ric_\tau(r,s)+E(r)] + \\ & 7(1+\frac{\epsilon}{2})\varsigma[(2^n-1)\log^+ T_H(r,s) + (2^{n-1}-1)(\log^+ Ric_\tau(r,s) + \end{aligned}$$

$$
\begin{aligned}
&\log^+ E(r))] + (1+\frac{\epsilon}{2})\varsigma[(2^n-1)\log^+ Y_{\mathbb{B}}(r) + 7\log^+\log^+ Y_{\mathbb{B}}(r)] + \\
&(1+\frac{\epsilon}{2})\varsigma[(15(2^{n-1}-2)+14+7)\log^+ r] + \ 2\log^+\log^+ r] \\
\underset{\cdot\cdot}{\leq}\ & 2^n T_H(r,s) + (2^n-1)[Ric_\tau(r,s) + E(r)] + \\
&7(1+\epsilon)\varsigma[(2^n-1)\log^+ T_H(r,s) + (2^{n-1}-1)(\log^+ Ric_\tau(r,s) + \\
&\log^+ E(r))] + (1+\epsilon)\varsigma[(2^n-1)\log^+ Y_{\mathbb{B}}(r) + \\
&(15(2^{n-1}-1)+7)\log^+ r].
\end{aligned}
$$

Notice that the coefficient of $\log^+ r$ is increased by one to handle the case when $Y_{\mathbb{B}}$ is bounded.

q.e.d.

As a corollary to this result,

$$
\begin{aligned}
\log^+ T_{D^nH}(r,s) \ \underset{\cdot\cdot}{\leq}\ & \log^+ T_H(r,s) + \log^+ Ric_\tau(r,s) + \log^+ E(r) + \\
& \log^+\log^+ Y_{\mathbb{B}}(r) + 2\log^+\log^+ r.
\end{aligned}
$$

Now, define for $0 < s < r$,

$$
\begin{aligned}
\mathbb{D}_n(r) &= \int_{M<r>} \log^+ \left|\frac{D^nH}{H}\right| \mathbf{Z}^n\sigma; \\
Q(r) &= \log^+ T_H(r,s) + \log^+ Ric_\tau(r,s) + \log^+ E(r) + \\
&\quad \log^+\log^+ Y_{\mathbb{B}}(r) + 2\log^+\log^+ r.
\end{aligned}
$$

Notice that for $r > 0$, $\mathbb{D}_n(r) \geq 0$.

THEOREM 2.9 [LEMMA OF THE HIGHER ORDER LOGARITHMIC DERIVATIVE] *Suppose assumptions* [A1],[A4],[A6],[A7],[A8] *are valid. Once again, let $H \not\equiv 0$ be a meromorphic function on M such that $dH \wedge \mathbb{B} \not\equiv 0$. Take $\epsilon > 0$, $0 < s < r$, and n in $\mathbb{N}$. Then,*

$$
\begin{aligned}
\int_{M<r>} \log^+\left|\frac{D^nH}{H}\right| \mathbf{Z}^n\sigma \ \underset{\cdot\cdot}{\leq}\ & n(1+\epsilon)\varsigma[7\log^+ T_H(r,s) + \log^+ Y_{\mathbb{B}}(r) + 6\log^+ r] + \\
& (n-1)(1+\epsilon)\varsigma[7\log^+ Ric_\tau(r,s) + 7\log^+ E(r) + \\
& 15\log^+\log^+ r]. \qquad (2.22)
\end{aligned}
$$

PROOF: If $D^nH \equiv 0$, then the estimate is trivial. Hence, suppose that $D^nH \not\equiv 0$. For $n = 1$, the estimate has been proven in 2.9. Additionally, 2.9 shows that

$$
\begin{aligned}
\int_{M<r>} \log^+\left|\frac{D(D^{n-1}H)}{D^{n-1}H}\right| \mathbf{Z}\sigma \ \underset{\cdot\cdot}{\leq}\ & (1+\frac{\epsilon}{2})\varsigma[7\log T_{D^{n-1}H}(r,s) + \\
& \log^+ Y_{\mathbb{B}}(r) + 6\log^+ r] \\
\underset{\cdot\cdot}{\leq}\ & (1+\frac{\epsilon}{2})\varsigma[7Q(r) + \log^+ Y_{\mathbb{B}}(r) + 6\log^+ r].
\end{aligned}
$$

Assume $n \geq 2$. Hence,

$$
\mathbb{D}_n(r) = \int_{M<r>} \log^+\left|\frac{D^nH}{H}\right| \mathbf{Z}^n\sigma
$$

$$\leq \int_{M<r>} \log^+ \left| \frac{D(D^{n-1})H}{D^{n-1}H} \right| \mathbf{Z}\sigma + \int_{M<r>} \log^+ \left| \frac{D^{n-1}H}{H} \right| \mathbf{Z}^{n-1}\sigma.$$

This implies that

$$\mathbb{D}_n(r) - \mathbb{D}_{n-1}(r) \underset{\cdot\cdot}{\leq} (1+\frac{\epsilon}{2})\varsigma[7Q(r) + \log^+ Y_{\mathbb{B}}(r) + 6\log^+ r].$$

Therefore,

$$\begin{aligned} \mathbb{D}_n(r) &= \left(\sum_{j=2}^{n}(\mathbb{D}_j(r) - \mathbb{D}_{j-1}(r)\right) + \mathbb{D}_1(r) \\ &\leq \mathbb{D}_1(r) + (n-1)(1+\frac{\epsilon}{2})\varsigma[7Q(r) + \log^+ Y_{\mathbb{B}}(r) + 6\log^+ r]. \end{aligned} \quad (2.23)$$

Now, by 2.21,

$$(1+\frac{\epsilon}{2})\varsigma[7\log^+\log^+ Y_{\mathbb{B}}(r) + \log^+ Y_{\mathbb{B}}(r)] \leq (1+\epsilon)\varsigma \log^+ Y_{\mathbb{B}}(r)$$

if $Y_{\mathbb{B}}$ is unbounded. Alternately, if $Y_{\mathbb{B}}$ is bounded, then

$$7\log^+\log^+ Y_{\mathbb{B}}(r) \leq \log^+\log^+ r.$$

Combining these two cases,

$$(1+\frac{\epsilon}{2})\varsigma[7\log^+\log^+ Y_{\mathbb{B}}(r) + \log^+ Y_{\mathbb{B}}(r)] \leq (1+\epsilon)\varsigma \log^+ Y_{\mathbb{B}}(r) + \log^+\log^+ r.$$

Using 2.9 to bound $\mathbb{D}_1(r)$ and expanding $Q(r)$ in 2.23,

$$\begin{aligned} \mathbb{D}_n(r) &\underset{\cdot\cdot}{\leq} (1+\epsilon)\varsigma[7\log T_H(r,s) + \log^+ Y_{\mathbb{B}}(r) + 6\log^+ r] + \\ &\quad (n-1)(1+\frac{\epsilon}{2})\varsigma[7\log^+ T_H(r,s) + 7\log^+ Ric_\tau(r,s) + 7\log^+ E(r) + \\ &\quad 7\log^+\log^+ Y_{\mathbb{B}}(r) + 14\log^+\log^+ r + \log^+ Y_{\mathbb{B}}(r) + 6\log^+ r] \\ &\underset{\cdot\cdot}{\leq} n(1+\epsilon)\varsigma[7\log T_H(r,s) + \log^+ Y_{\mathbb{B}}(r) + 6\log^+ r] + \\ &\quad (n-1)(1+\epsilon)\varsigma[7\log^+ Ric_\tau(r,s) + 7\log^+ E(r) + 15\log^+\log^+ r]. \end{aligned}$$

q.e.d.

Now, the notion of a coordinate function $f_{\mathfrak{a}\mathfrak{b}}$ needs to be dealt with to obtain the Lemma of the Logarithmic Derivative for the Wronski Determinant. Assume that [A1]-[A3] are valid. Take $\mathfrak{a}$ in V^* and $\mathfrak{b}$ in V_*^*. Denote $b = \mathbb{P}(\mathfrak{b})$. Assume as well that $f(M)$ is not contained in $\ddot{E}[b]$. Let $\mathfrak{v} : U \to V_*$ be a reduced representation of f. Then, $< \mathfrak{v}, \mathfrak{b} >$ is not identically zero. The coordinate function $f_{\mathfrak{a}\mathfrak{b}}$ is defined to be the unique meromorphic function on M such that

$$f_{\mathfrak{a}\mathfrak{b}}|U = \frac{< \mathfrak{v}, \mathfrak{a} >}{< \mathfrak{v}, \mathfrak{b} >} \quad (2.24)$$

for all possible choices of $\mathfrak{v}$.
This expression is unique. Suppose there exists a reduced representation of f denoted $\tilde{\mathfrak{v}} : \tilde{U} \to V$ such that $U \cap \tilde{U}$ is nonempty. Then, there exists a holomorphic function $g : U \cap \tilde{U} \to \mathbb{C}_*$ such that $\tilde{\mathfrak{v}} = g\mathfrak{v}$ on $U \cap \tilde{U}$. In turn, $\frac{<\tilde{\mathfrak{v}},\mathfrak{a}>}{<\tilde{\mathfrak{v}},\mathfrak{b}>} = \frac{<\mathfrak{v},\mathfrak{a}>}{<\mathfrak{v},\mathfrak{b}>}$ on $U \cap \tilde{U}$. Now, the characteristic functions of f and $f_{\mathfrak{a}\mathfrak{b}}$ will be related.

LEMMA 2.10 *Suppose* [A1]-[A3] *hold. Choose* $\mathfrak{a}$ *in* V^* *and* $\mathfrak{b}$ *in* V_*^* *such that* $f_{\mathfrak{a}\mathfrak{b}}$ *exists. Take* $s > 0$. *Then, there is a constant* $c(s)$ *such that for all* $r > s$,

$$T_{f_{\mathfrak{a}\mathfrak{b}}}(r, s) \leq T_f(r, s) + c(s). \tag{2.25}$$

PROOF: See Theorem 4.1 of Stoll [15].

Several important properties of the meromorphic differential operator D must now be considered.

LEMMA 2.11 *Assume that* [A1], [A4], *and* [A7] *are valid. Let* W *be a complex vector space. Let* U *be a nonempty, open subset of* M. *Let* $h : U \to \mathbb{C}$ *and* $\mathfrak{w} : U \to W$ *be holomorphic. Take* p *in* $\mathbb{Z}_+$. *Then,* 1. (*Leibniz Rule*

$$D^p(h\mathfrak{w}) \quad = \quad \sum_{q=0}^{p} \binom{p}{q} (D^q h)(D^{p-q}\mathfrak{w}).$$

2.

$$(h\mathfrak{w}) \wedge D(h\mathfrak{w}) \wedge D^2(h\mathfrak{w}) \wedge \ldots \wedge D^p(h\mathfrak{w}) = h^{p+1}\mathfrak{w} \wedge D\mathfrak{w} \wedge D^2\mathfrak{w} \wedge \ldots \wedge D^p\mathfrak{w}.$$

Next, suppose that $\mathfrak{z} = (\mathfrak{z}_1, \ldots, \mathfrak{z}_m) : \mathfrak{U} \to \mathfrak{U}'$ *is a chart on* M. *Denote*

$$\zeta \quad = \quad dz_1 \wedge \ldots \wedge dz_m.$$

Then, on U, *a holomorphic function* $\Theta_z \not\equiv 0$ *is given by* $\Theta|U = \Theta_z\zeta$. 3. *For* $p \geq 1$,

$$\Theta_z^{\frac{p(p+1)}{2}} D\mathfrak{w} \wedge D^2\mathfrak{w} \wedge \ldots \wedge D^p\mathfrak{w} \quad = \quad \mathfrak{w}' \wedge \mathfrak{w}'' \wedge \ldots \wedge \mathfrak{w}^{(p)}.] \tag{2.26}$$

4. *For* $p \geq 0$,

$$(h\mathfrak{w}) \wedge (h\mathfrak{w})' \wedge (h\mathfrak{w})'' \wedge \ldots \wedge (h\mathfrak{w})^{(p)} = h^{p+1}\mathfrak{w} \wedge \mathfrak{w}' \wedge \mathfrak{w}'' \wedge \ldots \wedge \mathfrak{w}^{(p)}.$$

5. *The identities 1. through 5. remain valid if* h *and* $\mathfrak{w}$ *are meromorphic.*

PROOF: 1. The Leibniz Rule is well known and can be shown by induction.

2. This identity can be proven by induction on p. It holds trivially when $p = 0$. Choose $p > 0$ and assume that the identity is valid for $p - 1$. Then,

$$(h\mathfrak{w}) \wedge D(h\mathfrak{w}) \wedge \ldots \wedge D^p(h\mathfrak{w}) = h^p\mathfrak{w} \wedge D\mathfrak{w} \wedge \ldots \wedge D^{p-1}\mathfrak{w} \wedge \sum_{q=0}^{p} \binom{p}{q} (D^q h)(D^{p-q}\mathfrak{w})$$

$$= h^{p+1}\mathfrak{w} \wedge D\mathfrak{w} \wedge \ldots \wedge D^p\mathfrak{w}.$$

3. 2.26 follows as well by induction on p. By definition,

$$(D\mathfrak{w})\Theta_z\zeta = (D\mathfrak{w})\Theta = d\mathfrak{w} \wedge \mathbb{B} = \mathfrak{w}'\zeta.$$

Therefore,

$$(D\mathfrak{w})\Theta_z = \mathfrak{w}' \quad \text{and} \quad D\mathfrak{w} = \frac{\mathfrak{w}'}{\Theta_z}.$$

Hence, 2.26 holds when $p = 1$. Assume that $p \geq 2$ and 2.26 is valid for $p-1$. Then,

$$\begin{aligned}\Theta_z^{\frac{p(p+1)}{2}} D\mathfrak{w} \wedge D^2\mathfrak{w} \wedge ... \wedge D^p\mathfrak{w} &= \Theta_z^p \left(\frac{\mathfrak{w}'}{\Theta_z}\right) \wedge \left(\Theta_z^{\frac{p(p-1)}{2}} D\left(\frac{\mathfrak{w}'}{\Theta_z}\right) \wedge ... \wedge D^{p-1}\left(\frac{\mathfrak{w}'}{\Theta_z}\right)\right) \\ &= \mathfrak{w}' \wedge \mathfrak{w}'' \wedge ... \wedge \mathfrak{w}^{(p)}.\end{aligned}$$

4. This Leibniz Formula can be shown directly by induction as done in 1. or can be derived from 3. in the following way:

$$\begin{aligned}(h\mathfrak{w}) \wedge (h\mathfrak{w})' \wedge (h\mathfrak{w})'' \wedge ... \wedge (h\mathfrak{w})^{(p)} &= \Theta_z^{\frac{p(p+1)}{2}} h\mathfrak{w} \wedge D(h\mathfrak{w}) \wedge ... \wedge D^p(h\mathfrak{w}) \\ &= h^{p+1}\Theta_z^{\frac{p(p+1)}{2}} \mathfrak{w} \wedge D\mathfrak{w} \wedge ... \wedge D^p\mathfrak{w} \\ &= h^{p+1}\mathfrak{w} \wedge \mathfrak{w}' \wedge \mathfrak{w}'' \wedge ... \wedge \mathfrak{w}^{(p)}.\end{aligned}$$

5. If $\mathfrak{w}$ is meromorphic, then let P be the set of inessential singularities of $\mathfrak{w}$ in U. Then, $\mathfrak{w}|(U-P)$ is holomorphic and the above identities are valid on $U-P$. It must be shown that the respective vector functions extend to meromorphic functions on U. Take a in P. Then, there exists an open, connected neighborhood U_0 of a such that there exists a chart $\mathfrak{z} = (z_1, ..., z_m) : U_0 \to U_0'$ on M and such that there exist holomorphic $0 \not\equiv g : U_0 \to \mathbb{C}$ and $\mathfrak{v} : U_0 \to W$ with $g\mathfrak{w} = \mathfrak{v}$. Then, g' and $\mathfrak{v}'$ are holomorphic on U_0. Additionally,

$$\begin{aligned}D\mathfrak{w} &= D\left(\frac{\mathfrak{v}}{g}\right) \\ &= \frac{gD\mathfrak{v} - (Dg)\mathfrak{v}}{g^2} \\ &= \frac{g\mathfrak{v}' - g'\mathfrak{v}}{g^2\Theta_z}.\end{aligned}$$

Therefore, $D\mathfrak{w}$ is meromorphic on U_0. Since this holds for each a in P, then the vector function $D\mathfrak{w}$ is meromorphic on U. By induction, $D^q\mathfrak{w}$ is a meromorphic vector function on U for all q in $\mathbb{N}$. Thus, $\mathfrak{w} \wedge D\mathfrak{w} \wedge ... \wedge D^p\mathfrak{w}$ and $D\mathfrak{w} \wedge ... \wedge D^p\mathfrak{w}$ are meromorphic. Similarly,

$$\mathfrak{w}' = \frac{g\mathfrak{v}' - g'\mathfrak{v}}{g^2}$$

is meromorphic on U. Consequently, by induction, $\mathfrak{w}^{(q)}$ is meromorphic on U for all q in $\mathbb{N}$. Thus, $\mathfrak{w} \wedge \mathfrak{w}' \wedge ... \wedge \mathfrak{w}^{(p)}$ and $\mathfrak{w}' \wedge ... \wedge \mathfrak{w}^{(p)}$ are meromorphic on U. q.e.d.

Next, suppose that assumptions [A1] through [A8] are valid. Take a family $\{U_\lambda, \mathfrak{z}_\lambda, \mathfrak{v}_\lambda\}_{\lambda \in \Lambda}$ satisfying the following properties:
1. For λ in Λ, U_λ is a nonempty, open, connected subset of M.
2. For λ in Λ, $\mathfrak{z}_\lambda = (z_{1\lambda}, ..., z_{m\lambda}) : U_\lambda \to U_\lambda'$ is a chart on M. On U_λ, define

$$\zeta_\lambda = dz_{1\lambda} \wedge ... \wedge dz_{m\lambda}.$$

3. For λ in Λ, $\mathfrak{v}_\lambda : U_\lambda \to V$ is a reduced representation of f.
4. $\mathfrak{U} = \{U_\lambda\}_{\lambda\in\Lambda}$ is a covering of M. For $\lambda = (\lambda_0, ..., \lambda_p)$ in Λ^{p+1},

$$U_\lambda = U_{\lambda_0} \cap ... \cap U_{\lambda_p} \qquad \text{and} \qquad \Lambda[p] = \{\lambda \in \Lambda^{p+1} | U_\lambda \neq \emptyset\}.$$

5. For every (λ, μ) in $\Lambda[1]$, there exist unique holomorphic functions $\Delta_{\lambda\mu}$ and $g_{\lambda\mu}$ such that on $U_{\lambda\mu}$,

$$\zeta_\lambda = \Delta_{\lambda\mu}\zeta_\mu \qquad \text{and} \qquad \mathfrak{v}_\lambda = g_{\lambda\mu}\mathfrak{v}_\mu.$$

Then,

$$\begin{array}{rllll} \Delta_{\lambda\lambda} = 1 & = g_{\lambda\lambda} & \text{on } U_\lambda & \text{if } \lambda \in \Lambda; \\ \Delta_{\lambda\mu}\Delta_{\mu\lambda} = 1 & = g_{\lambda\mu}g_{\mu\lambda} & \text{on } U_{\lambda\mu} & \text{if } (\lambda, \mu) \in \Lambda[1]; \\ \Delta_{\lambda\mu}\Delta_{\mu\rho}\Delta_{\rho\lambda} = 1 & = g_{\lambda\mu}g_{\mu\rho}g_{\rho\lambda} & \text{on } U_{\lambda\mu\rho} & \text{if } (\lambda, \mu, \rho) \in \Lambda[2]. \end{array}$$

Therefore, this yields the basic cocycles $\{\Delta_{\lambda\mu}\}_{(\lambda,\mu)\in\Lambda[1]}$ for the canonical bundle K on M and $\{g_{\lambda\mu}\}_{(\lambda,\mu)\in\Lambda[1]}$ for the hyperplane section bundle L_f of f.

6. For each λ in Λ, there exists a holomorphic frame $\overset{\triangle}{\mathfrak{v}_\lambda}$ of L_f over U_λ such that on $U_{\lambda\mu}$,

$$\overset{\triangle}{\mathfrak{v}_\mu} \quad = \quad g_{\lambda\mu}\overset{\triangle}{\mathfrak{v}_\lambda} \ .$$

Denote $\Theta_\lambda = \Theta_{\mathfrak{z}_\lambda}$. Then, on $U_{\lambda\mu}$,

$$\Theta_\mu\zeta_\mu = \Theta = \Theta_\lambda\zeta_\lambda = \Theta_\lambda\Delta_{\lambda\mu}\zeta_\mu.$$

Hence,

$$\Theta_\mu \quad = \quad \Theta_\lambda\Delta_{\lambda\mu}.$$

Take p in $\mathbb{N}$. Then,

$$\begin{aligned} \mathfrak{v}_\lambda \wedge D\mathfrak{v}_\lambda \wedge ... \wedge D^p\mathfrak{v}_\lambda \quad &= \quad g_{\lambda\mu}\mathfrak{v}_\mu \wedge D(g_{\lambda\mu}\mathfrak{v}_\mu) \wedge ... \wedge D^p(g_{\lambda\mu}\mathfrak{v}_\mu) \\ &= \quad (g_{\lambda\mu})^{p+1}\mathfrak{v}_\mu \wedge D\mathfrak{v}_\mu \wedge ... \wedge D^p\mathfrak{v}_\mu. \end{aligned}$$

Therefore,

$$(\Delta_{\lambda\mu}\Theta_\lambda)^{\frac{p(p+1)}{2}}\mathfrak{v}_\lambda \wedge D\mathfrak{v}_\lambda \wedge ... \wedge D^p\mathfrak{v}_\lambda = (g_{\lambda\mu})^{p+1}\Theta_\mu^{\frac{p(p+1)}{2}}\ \mathfrak{v}_\mu \wedge D\mathfrak{v}_\mu \wedge ... \wedge D^p\mathfrak{v}_\mu \ .$$

and consequently

$$(\Delta_{\lambda\mu})^{\frac{p(p+1)}{2}}\mathfrak{v}_\lambda \wedge \mathfrak{v}'_\lambda \wedge ... \wedge \mathfrak{v}_\lambda^{(p)} = (g_{\lambda\mu})^{p+1}\mathfrak{v}_\mu \wedge \mathfrak{v}'_\mu \wedge ... \wedge \mathfrak{v}_\mu^{(p)}.$$

Notice that on the left hand side of this equation, the differentiation is done with respect to $\mathfrak{z}_\lambda$ while on the right it is with respect to $\mathfrak{z}_\mu$. This equation can be rewritten as

$$\mathfrak{v}_\lambda \wedge \mathfrak{v}'_\lambda \wedge ... \wedge \mathfrak{v}^{(p)}_\lambda = (g_{\lambda\mu})^{p+1}(\Delta_{\mu\lambda})^{\frac{p(p+1)}{2}} \mathfrak{v}_\mu \wedge \mathfrak{v}'_\mu \wedge ... \wedge \mathfrak{v}^{(p)}_\mu .$$

Denote $\mathfrak{v}_{\lambda\underline{p}} = \mathfrak{v}_\lambda \wedge \mathfrak{v}'_\lambda \wedge ... \wedge \mathfrak{v}^{(p)}_\lambda$. Then, on $U_{\lambda\mu}$ for (λ, μ) in $\Lambda[1]$,

$$\mathfrak{v}_{\lambda\underline{p}} = (g_{\lambda\mu})^{p+1}(\Delta_{\mu\lambda})^{\frac{p(p+1)}{2}} \mathfrak{v}_{\mu\underline{p}}.$$

Let $(\underset{p+1}{\wedge} V)_M = M\times \underset{p+1}{\wedge} V$ be the] trivial bundle. A holomorphic section $\tilde{\mathfrak{v}}_{\lambda\underline{p}} : U_\lambda \to (\underset{p+1}{\wedge} V)_M$ is given by $\tilde{\mathfrak{v}}_{\lambda\underline{p}}(x) = (x, \mathfrak{v}_{\lambda\underline{p}}(x))$ for all x in U_λ. On $U_{\lambda\mu}$ for (λ, μ) in $\Lambda[1]$,

$$\begin{aligned} \tilde{\mathfrak{v}}_{\lambda\underline{p}} &= (g_{\lambda\mu})^{p+1}(\Delta_{\mu\lambda})^{\frac{p(p+1)}{2}} \tilde{\mathfrak{v}}_{\mu\underline{p}}; \\ \tilde{\mathfrak{v}}_{\lambda\underline{p}} \otimes (\overset{\triangle}{\mathfrak{v}_\lambda})^{p+1} \otimes (\zeta_\lambda)^{\frac{p(p+1)}{2}} &= \tilde{\mathfrak{v}}_{\mu\underline{p}} \otimes (g_{\lambda\mu}\overset{\triangle}{\mathfrak{v}_\lambda})^{p+1} \otimes (\Delta_{\mu\lambda}\zeta_\lambda)^{\frac{p(p+1)}{2}}; \\ &= \tilde{\mathfrak{v}}_{\mu\underline{p}} \otimes (\overset{\triangle}{\mathfrak{v}_\mu})^{p+1} \otimes (\zeta_\mu)^{\frac{p(p+1)}{2}}. \end{aligned}$$

Hence, there exists a unique holomorphic section F_p over M of the bundle

$$(\underset{p+1}{\wedge} V)_M \otimes L_f^{p+1} \otimes K^{\frac{p(p+1)}{2}}$$

such that for all λ in Λ,

$$F_p|U_\lambda = \tilde{\mathfrak{v}}_{\lambda\underline{p}} \otimes (\overset{\triangle}{\mathfrak{v}_\lambda})^{p+1} \otimes (\zeta_\lambda)^{\frac{p(p+1)}{2}}.$$

Here, F_p is called the <u>canonical representation section</u> of order p. Take $p = n$. Then, the resulting holomorphic line bundle is

$$(\underset{n+1}{\wedge} V)_M \otimes L_f^{n+1} \otimes K^{\frac{n(n+1)}{2}}.$$

Due to assumption [A5], $\mathfrak{v}_{\lambda\underline{n}} \not\equiv 0$. Thus, $\tilde{\mathfrak{v}}_{\lambda\underline{n}} \not\equiv 0$. Hence, $F_n \not\equiv 0$.

Let $\mathfrak{b} = (\mathfrak{b}_0, ..., \mathfrak{b}_n)$ be a base of V^*. Let $\mathfrak{e} = (\mathfrak{e}_0, ..., \mathfrak{e}_n)$ be the dual base to $\mathfrak{b}$ in V. Since f is general for $\mathbb{B}$, the map f is linearly non- degenerate. Therefore, the coordinate functions $f_j = f_{\mathfrak{b}_j,\mathfrak{b}_0}$ are defined and $f_j \not\equiv 0$ for $0 \leq j \leq n$. Each f_j is a meromorphic function on M and $f_0 = 1$. Thus, a meromorphic vector function $\mathfrak{w} : M \to V$ is defined by

$$\mathfrak{w} = f_0\mathfrak{e}_0 + ... + f_n\mathfrak{e}_n \not\equiv 0.$$

Then, a meromorphic function $W_{f,\mathfrak{b}}$ called the <u>Wronski determinant</u> of f for $\mathfrak{b}$ is given by

$$W_{f,\mathfrak{b}} = \frac{< \mathfrak{w} \wedge D\mathfrak{w} \wedge ... \wedge D^n\mathfrak{w}, \mathfrak{b}_0 \wedge \mathfrak{b}_1 \wedge ... \wedge \mathfrak{b}_n >}{f_0 f_1 \cdots f_n}.$$

Choose any λ in Λ. $\mathfrak{w}$ and $W_{f,\mathfrak{b}}$ will now be calculated on U_λ. Since $\mathfrak{b}$ is the dual base of $\mathfrak{e}$,

$$\mathfrak{v}_\lambda = < \mathfrak{v}_\lambda, \mathfrak{b}_0 > \mathfrak{e}_0 + ...+ < \mathfrak{v}_\lambda, \mathfrak{b}_n > \mathfrak{e}_n.$$

Consequently,

$$\begin{aligned}\frac{\mathfrak{v}_\lambda}{<\mathfrak{v}_\lambda,\mathfrak{b}_0>} &= \mathfrak{e}_0+\frac{<\mathfrak{v}_\lambda,\mathfrak{b}_1>}{<\mathfrak{v}_\lambda,\mathfrak{b}_0>}\mathfrak{e}_1+\ldots+\frac{<\mathfrak{v}_\lambda,\mathfrak{b}_n>}{<\mathfrak{v}_\lambda,\mathfrak{b}_0>}\mathfrak{e}_n\\ &= \mathfrak{e}_0+(f_{\mathfrak{b}_1,\mathfrak{b}_0}|U_\lambda)\mathfrak{e}_1+\ldots+(f_{\mathfrak{b}_n,\mathfrak{b}_0}|U_\lambda)\mathfrak{e}_n\\ &= (\mathfrak{e}_0+f_1\mathfrak{e}_1+\ldots+f_n\mathfrak{e}_n)|U_\lambda\\ &= \mathfrak{w}|U_\lambda.\end{aligned}$$

Hence,

$$\begin{aligned}\mathfrak{w}\wedge D\mathfrak{w}\wedge\ldots\wedge D^n\mathfrak{w}|U_\lambda &= \frac{\mathfrak{v}_\lambda}{<\mathfrak{v}_\lambda,\mathfrak{b}_0>}\wedge D\left(\frac{\mathfrak{v}_\lambda}{<\mathfrak{v}_\lambda,\mathfrak{b}_0>}\right)\wedge\ldots\wedge D^n\left(\frac{\mathfrak{v}_\lambda}{<\mathfrak{v}_\lambda,\mathfrak{b}_0>}\right)\\ &= \frac{1}{<\mathfrak{v}_\lambda,\mathfrak{b}_0>^{n+1}}(\mathfrak{v}_\lambda\wedge D\mathfrak{v}_\lambda\wedge\ldots\wedge D^n\mathfrak{v}_\lambda)\\ &= \frac{\mathfrak{v}_{\lambda\underline{n}}}{\Theta_\lambda^{\frac{n(n+1)}{2}}<\mathfrak{v}_\lambda,\mathfrak{b}_0>^{n+1}}\not\equiv 0.\end{aligned}$$

In particular,

$$\mathfrak{w}\wedge D\mathfrak{w}\wedge\ldots\wedge D^n\mathfrak{w}|U_\lambda\not\equiv 0.$$

In addition,

$$\begin{aligned}W_{f,\mathfrak{b}}|U_\lambda &= \frac{<\mathfrak{v}_\lambda\wedge D\mathfrak{v}_\lambda\wedge\ldots\wedge D^n\mathfrak{v}_\lambda,\mathfrak{b}_0\wedge\mathfrak{b}_1\wedge\ldots\wedge\mathfrak{b}_n>}{<\mathfrak{v}_\lambda,\mathfrak{b}_0><\mathfrak{v}_\lambda,\mathfrak{b}_1>\cdots<\mathfrak{v}_\lambda,\mathfrak{b}_n>}\\ &= \frac{<\mathfrak{v}_{\lambda\underline{n}},\mathfrak{b}_0\wedge\mathfrak{b}_1\wedge\ldots\wedge\mathfrak{b}_n>}{\Theta_\lambda^{\frac{n(n+1)}{2}}<\mathfrak{v}_\lambda,\mathfrak{b}_0><\mathfrak{v}_\lambda,\mathfrak{b}_1>\cdots<\mathfrak{v}_\lambda,\mathfrak{b}_n>}.\end{aligned}$$

Define $b_j=\mathbb{P}(\mathfrak{b}_j)$ in $\mathbb{P}(V^*)$ for $0\le j\le n$. Recall that by definition $\beta=\mu_\Theta$. Then,

$$\begin{aligned}\mu_{F_n}|U_\lambda &= \mu_{\mathfrak{v}_{\lambda\underline{n}}}=\mu_{<\mathfrak{v}_{\lambda\underline{n}},\mathfrak{b}_0\wedge\mathfrak{b}_1\wedge\ldots\wedge\mathfrak{b}_n>};\\ \beta|U_\lambda &= \mu_{\Theta_\lambda};\\ \mu_{f,b_j}|U_\lambda &= \mu_{<\mathfrak{v}_\lambda,\mathfrak{b}_j>}.\end{aligned}$$

Recall that the divisor of the meromorphic function $W_{f,\mathfrak{b}}$ is given by

$$\mu_{W_{f,\mathfrak{b}}} = \mu_{W_{f,\mathfrak{b}},0}-\mu_{W_{f,\mathfrak{b}},\infty}.$$

Therefore, for all λ in Λ,

$$\mu_{W_{f,\mathfrak{b}}}|U_\lambda = \mu_{F_n}|U_\lambda-\frac{n(n+1)}{2}\beta|U_\lambda-\sum_{j=0}^{n}\mu_{f,b_j}|U_\lambda.$$

Consequently,

$$\mu_{W_{f,\mathfrak{b}}} = \mu_{F_n}-\frac{n(n+1)}{2}\beta-\sum_{j=0}^{n}\mu_{f,b_j}.$$

Thus, for $0 < s < r$,

$$N_{W_{f,\mathfrak{b}}}(r,s) = N_{F_n}(r,s) - \frac{n(n+1)}{2} N_\beta(r,s) - \sum_{j=0}^{n} N_{f,\mathfrak{b}_j}(r,s). \quad (2.27)$$

Now, the following important result can be proven.

THEOREM 2.12 [LEMMA OF THE LOGARITHMIC DERIVATIVE FOR THE WRONSKI DETERMINANT] *Suppose* [A1]-[A8] *hold. Let* $\mathfrak{b} = (\mathfrak{b}_0, ..., \mathfrak{b}_n)$ *be a base of* V^*. *Take* $\epsilon > 0$ *and* $0 < s < r$. *Then,*

$$\begin{aligned}\int_{M\langle r\rangle} \log^+ |W_{f,\mathfrak{b}}| \mathbf{Z}^{\frac{n(n+1)}{2}} \sigma \;\underset{\because}{\leq}\; & \frac{1}{2} n^2(n+1)(1+\epsilon)\varsigma[7\log^+ T_f(r,s) + \\ & \log^+ Y_{\mathbb{B}}(r) + 7\log^+ r] + \\ & \frac{7}{2} n^2(n-1)(1+\epsilon)\varsigma[\log^+ Ric_\tau(r,s) + \\ & \log^+ E(r)]. \end{aligned} \quad (2.28)$$

PROOF: Define $\mathfrak{e} = (\mathfrak{e}_0, ..., \mathfrak{e}_n)$, f_j, and $\mathfrak{w}$ as before. Then,

$$\begin{aligned} \mathfrak{w} &= \mathfrak{e}_0 + f_1\mathfrak{e}_1 + ... + f_n\mathfrak{e}_n \\ D\mathfrak{w} &= Df_1\mathfrak{e}_1 + ... + Df_n\mathfrak{e}_n \\ &\;\cdot \\ &\;\cdot \\ &\;\cdot \\ D^n\mathfrak{w} &= D^n f_1\mathfrak{e}_1 + ... + D^n f_n\mathfrak{e}_n \end{aligned}$$

which thus yields

$$\mathfrak{w} \wedge D\mathfrak{w} \wedge ... \wedge D^n\mathfrak{w} = \begin{vmatrix} 1 & f_1 & \cdot & \cdot & \cdot & f_n \\ 0 & Df_1 & \cdot & \cdot & \cdot & Df_n \\ \cdot & & & & & \cdot \\ \cdot & & & & & \cdot \\ \cdot & & & & & \cdot \\ 0 & D^n f_1 & \cdot & \cdot & \cdot & D^n f_n \end{vmatrix} \mathfrak{e}_0 \wedge \mathfrak{e}_1 \wedge ... \wedge \mathfrak{e}_n.$$

Since $f_0 = 1$,

$$\begin{aligned} W_{f,\mathfrak{b}} &= \frac{< \mathfrak{w} \wedge D\mathfrak{w} \wedge ... \wedge D^n\mathfrak{w}, \mathfrak{b}_0 \wedge \mathfrak{b}_1 \wedge ... \wedge \mathfrak{b}_n >}{f_0 f_1 \cdots f_n} \\ &= \frac{\begin{vmatrix} 1 & f_1 & \cdot & \cdot & \cdot & f_n \\ 0 & Df_1 & \cdot & \cdot & \cdot & Df_n \\ \cdot & & & & & \cdot \\ \cdot & & & & & \cdot \\ \cdot & & & & & \cdot \\ 0 & D^n f_1 & \cdot & \cdot & \cdot & D^n f_n \end{vmatrix}}{f_0 f_1 \cdots f_n} \end{aligned}$$

$$= \frac{\begin{vmatrix} Df_1 & \cdots & Df_n \\ \vdots & & \vdots \\ D^n f_1 & \cdots & D^n f_n \end{vmatrix}}{f_1 \cdots f_n}.$$

In addition,

$$\begin{aligned}
|W_{f,\mathfrak{b}}|\,\mathbf{Z}^{\frac{n(n+1)}{2}} &= \mathbf{Z}^{\frac{n(n+1)}{2}} \frac{1}{|f_1 \cdots f_n|} \begin{vmatrix} Df_1 & \cdots & Df_n \\ \vdots & & \vdots \\ D^n f_1 & \cdots & D^n f_n \end{vmatrix} \\
&= \begin{vmatrix} \mathbf{Z}\frac{Df_1}{f_1} & \cdots & \mathbf{Z}\frac{Df_n}{f_n} \\ \vdots & & \vdots \\ \mathbf{Z}^n \frac{D^n f_1}{f_1} & \cdots & \mathbf{Z}^n \frac{D^n f_n}{f_n} \end{vmatrix} \\
&\leq \prod_{j=1}^{n} \left(\sqrt{\sum_{k=1}^{n} \mathbf{Z}^k \left| \frac{D^k f_j}{f_j} \right|^2} \right).
\end{aligned}$$

Then,

$$\begin{aligned}
\log^+ |W_{f,\mathfrak{b}}|\,\mathbf{Z}^{\frac{n(n+1)}{2}} &\leq \sum_{j=1}^{n} \frac{1}{2} \log^+ \left(\sum_{k=1}^{n} \mathbf{Z}^k \left| \frac{D^k f_j}{f_j} \right|^2 \right) \\
&\leq \sum_{j=1}^{n} \sum_{k=1}^{n} \log^+ \mathbf{Z}^k \left| \frac{D^k f_j}{f_j} \right| + \frac{n \log n}{2}.
\end{aligned}$$

Thus, by first using 2.22 and then 2.25,

$$\begin{aligned}
\int_{M<r>} \log^+ |W_{f,\mathfrak{b}}|\,\mathbf{Z}^{\frac{n(n+1)}{2}} \sigma &\leq \sum_{j=1}^{n} \sum_{k=1}^{n} \int_{M<r>} \log^+ \mathbf{Z}^k \left| \frac{D^k f_j}{f_j} \right| \sigma + \frac{1}{2} n \varsigma \log n \\
&\mathrel{\underset{\cdot\cdot}{\leq}} \sum_{j=1}^{n} \sum_{k=1}^{n} \{\, k(1+\frac{\epsilon}{2})\varsigma[7 \log^+ T_{f_j}(r,s) + \\
&\quad \log^+ Y_{\mathbb{B}}(r) + 6 \log^+ r] + \\
&\quad 7(k-1)(1+\frac{\epsilon}{2})\varsigma[\log^+ Ric_\tau(r,s) + \log^+ E(r)] + \\
&\quad 15(k-1)(1+\frac{\epsilon}{2})\varsigma \log^+ \log^+ r \,\} + \frac{1}{2} n \varsigma \log n \\
&\mathrel{\underset{\cdot\cdot}{\leq}} n \sum_{k=1}^{n} \{\, k(1+\frac{\epsilon}{2})\varsigma[7 \log^+ T_f(r,s) + \\
&\quad \log^+ Y_{\mathbb{B}}(r) + 6 \log^+ r] + \\
&\quad 7(k-1)(1+\frac{\epsilon}{2})\varsigma[\log^+ Ric_\tau(r,s) + \log^+ E(r)] +
\end{aligned}$$

$$
\begin{aligned}
& 15(k-1)(1+\frac{\epsilon}{2})\varsigma \log^+ \log^+ r \;\} + \frac{1}{2} n\varsigma \log n \\
= \;& n\left(\frac{n(n+1)}{2}\right)(1+\frac{\epsilon}{2})\varsigma[7\log^+ T_f(r,s) + \\
& \log^+ Y_{\mathbb{B}}(r) + 6\log^+ r] + \\
& 7n\frac{(n-1)n}{2}(1+\frac{\epsilon}{2})\varsigma[\log^+ Ric_\tau(r,s) + \log^+ E(r)] + \\
& 15n\frac{(n-1)n}{2}(1+\frac{\epsilon}{2})\varsigma \log^+ \log^+ r + \frac{1}{2} n\varsigma \log n.
\end{aligned}
$$

Then, applying 2.21,

$$
\begin{aligned}
\int_{M<r>} \log^+ |W_{f,\mathfrak{b}}| \, \mathbf{Z}^{\frac{n(n+1)}{2}} \sigma \;\; \underset{\because}{\leq} \;\; & \frac{1}{2} n^2 (n+1)(1+\epsilon)\varsigma[7\log^+ T_f(r,s) + \\
& \log^+ Y_{\mathbb{B}}(r) + 6\log^+ r] + \\
& \frac{7}{2} n^2 (n-1)(1+\epsilon)\varsigma[\log^+ Ric_\tau(r,s) + \log^+ E(r)].
\end{aligned}
$$

q.e.d.

Next, the Second Main Theorem for Fixed Targets for meromorphic maps $f : M \to \mathbb{P}(V)$ must be proven. The Lemma of the Logarithmic Derivative is used to derive this result by the Cartan method. Three additional assumptions must be made.

(A9) G is a finite subset of $\mathbb{P}(V^*)$ such that $n+1 \leq q = \#G < \infty$.
(A10) G is in general position.
(A11) f is linearly non-degenerated.

Here, G is in general position if and only if every subset B of G such that $\#B = n+1$ is not contained in any proper linear projective subspace of $\mathbb{P}(V^*)$. Also, f is called linearly non-degenerated if $f(M)$ is not contained in the hyperplane $\ddot{E}[a]$ for any a in $\mathbb{P}(V^*)$.

Define the space $\mathfrak{B} = \{B | B \subseteq G \text{ and } \#B = n+1\}$. Then, for every B in $\mathfrak{B}$, define the number $\square B \square$ as follows. First, enumerate $B = \{b_0, b_1, ..., b_n\}$. Then, taking $\mathfrak{b}_j$ in V^*_* such that $\mathbb{P}(\mathfrak{b}_j) = b_j$, denote

$$
0 < \square B \square = \frac{\| \mathfrak{b}_0 \wedge ... \wedge \mathfrak{b}_n \|}{\| \mathfrak{b}_0 \| \cdots \| \mathfrak{b}_n \|} \leq 1. \tag{2.29}
$$

$\square B \square$ does not depend upon the choice of the $\mathfrak{b}_j$ and the enumeration of B itself. Also, define the number $\Gamma(G)$ to be

$$
0 < \Gamma(G) \;=\; \min\{\square B \square | B \in \mathfrak{B}\} \leq 1.
$$

Lemma 2.13 [Product to Sum Estimate] *For x in $\mathbb{P}(V)$, there exists B in $\mathfrak{B}$ which depends on x such that*

$$
\prod_{g \in G} \frac{1}{\square x, g \square} \leq \left(\frac{n+1}{\Gamma(G)}\right)^{q-n-1} \prod_{g \in B} \frac{1}{\square x, g \square}. \tag{2.30}
$$

PROOF: See Lemma 2.2.1 of Bardis [2].

Then, this estimate is used to prove the following result.

THEOREM 2.14 [POINTWISE SECOND MAIN THEOREM] *Suppose* [A1]-[A11] *are valid. Let* $q = \#G$. *Define* $p = \#\mathfrak{B} = \binom{q}{n+1}$. *Let* $\mathfrak{a} = (\mathfrak{a}_0, ..., \mathfrak{a}_n)$ *be an orthonormal base for* V^*. *Abbreviate* $W = W_{f,\mathfrak{a}}$ *and* $a_j = \mathbb{P}(\mathfrak{a}_j)$. *For every* B *in* $\mathfrak{B}$, *select an enumeration* $B = \{b_0, ..., b_n\}$. *Then, choose* $\mathfrak{b}_j$ *in* V^* *such that* $\| \mathfrak{b}_j \| = 1$ *and* $b_j = \mathbb{P}(\mathfrak{b}_j)$. *Abbreviate* $W_B = W_{f,\mathfrak{b}}$ *where* $\mathfrak{b} = \{\mathfrak{b}_0, ..., \mathfrak{b}_n\}$ *is a base of* V^* *defined by* B. *Note that* $|W_B|$ *does not depend on these choices. Then,*

$$\sum_{g\in G} \log \frac{1}{\square f, g\square} \leq (q-n)\log\frac{n+1}{\Gamma(G)} + \log\frac{1}{|W|\mathbf{Z}^{\frac{n(n+1)}{2}}} + \sum_{j=0}^{n}\log\frac{1}{\square f, a_j\square} + \sum_{B\in\mathfrak{B}} \log^+ |W_B|\mathbf{Z}^{\frac{n(n+1)}{2}} + \binom{q}{n+1}. \tag{2.31}$$

PROOF: Let $\{U_\lambda, \mathfrak{v}_\lambda, \mathfrak{z}_\lambda\}$ be a representation atlas of f. Take B in $\mathfrak{B}$. For $0 \leq j \leq n$, take $\mathfrak{b}_j$ as indicated above. Then, on U_λ,

$$W_B = \frac{< \mathfrak{v}_\lambda \wedge D\mathfrak{v}_\lambda \wedge ... \wedge D^n\mathfrak{v}_\lambda, \mathfrak{b}_0 \wedge \mathfrak{b}_1 \wedge ... \wedge \mathfrak{b}_n >}{< \mathfrak{v}_\lambda, \mathfrak{b}_0 >< \mathfrak{v}_\lambda, \mathfrak{b}_1 > \cdots < \mathfrak{v}_\lambda, \mathfrak{b}_n >};$$

$$W = \frac{< \mathfrak{v}_\lambda \wedge D\mathfrak{v}_\lambda \wedge ... \wedge D^n\mathfrak{v}_\lambda, \mathfrak{a}_0 \wedge \mathfrak{a}_1 \wedge ... \wedge \mathfrak{a}_n >}{< \mathfrak{v}_\lambda, \mathfrak{a}_0 >< \mathfrak{v}_\lambda, \mathfrak{a}_1 > \cdots < \mathfrak{v}_\lambda, \mathfrak{a}_n >}.$$

Also, $\mathfrak{b}_0 \wedge ... \wedge \mathfrak{b}_n = \gamma_B \mathfrak{a}_0 \wedge ... \wedge \mathfrak{a}_n$ where $|\gamma_B| = \square B \square$ since $\| \mathfrak{a}_0 \wedge ... \wedge \mathfrak{a}_n \| = 1 = \| \mathfrak{b}_j \|$ for all $0 \leq j \leq n$. In turn, this implies that

$$\frac{W_B}{W} = \gamma_B \frac{< \mathfrak{v}_\lambda, \mathfrak{a}_0 >< \mathfrak{v}_\lambda, \mathfrak{a}_1 > \cdots < \mathfrak{v}_\lambda, \mathfrak{a}_n >}{< \mathfrak{v}_\lambda, \mathfrak{b}_0 >< \mathfrak{v}_\lambda, \mathfrak{b}_1 > \cdots < \mathfrak{v}_\lambda, \mathfrak{b}_n >};$$

$$\frac{|W_B|}{|W|} = \square B \square \prod_{j=0}^{n} \frac{\| \mathfrak{v}_\lambda \| \| \mathfrak{b}_j \| \, | < \mathfrak{v}_\lambda, \mathfrak{a}_j > |}{| < \mathfrak{v}_\lambda, \mathfrak{b}_j > | \, \| \mathfrak{v}_\lambda \| \| \mathfrak{a}_j \|} = \square B \square \prod_{j=0}^{n} \frac{\square f, a_j \square}{\square f, b_j \square}.$$

Therefore,

$$\prod_{g\in B} \frac{1}{\square f, g \square} = \frac{|W_B|}{|W|} \frac{1}{\square B \square} \prod_{j=0}^{n} \frac{1}{\square f, a_j \square} = \frac{\mathbf{Z}^{\frac{n(n+1)}{2}} |W_B|}{\mathbf{Z}^{\frac{n(n+1)}{2}} |W|} \frac{1}{\square B \square} \prod_{j=0}^{n} \frac{1}{\square f, a_j \square}.$$

Then, by 2.30,

$$\prod_{g\in G} \frac{1}{\square f, g \square} \leq \left(\frac{n+1}{\Gamma(G)}\right)^{q-n-1} \sum_{B\in\mathfrak{B}} \left(\frac{|W_B|\mathbf{Z}^{\frac{n(n+1)}{2}}}{|W|\mathbf{Z}^{\frac{n(n+1)}{2}}} \frac{1}{\Gamma(G)} \prod_{j=0}^{n} \frac{1}{\square f, a_j \square} \right)$$

$$= \left(\frac{n+1}{\Gamma(G)}\right)^{q-n-1} \frac{1}{|W|\mathbf{Z}^{\frac{n(n+1)}{2}}} \frac{1}{\Gamma(G)} \prod_{j=0}^{n} \frac{1}{\square f, a_j \square} \left(\sum_{B\in\mathfrak{B}} |W_B|\mathbf{Z}^{\frac{n(n+1)}{2}} \right).$$

Finally, using 2.14 and since $\#\mathfrak{B} = \binom{q}{n+1}$,

$$\begin{aligned}
\log \prod_{g \in G} \frac{1}{\|f, g\|} &= \sum_{g \in G} \log \frac{1}{\|f, g\|} \\
&\leq (q-n) \log \frac{n+1}{\Gamma(G)} + \log \frac{1}{|W| \mathbf{Z}^{\frac{n(n+1)}{2}}} + \sum_{j-0}^{n} \log \frac{1}{\|f, a_j\|} + \\
&\quad \sum_{B \in \mathfrak{B}} \log^{+} |W_B| \mathbf{Z}^{\frac{n(n+1)}{2}} + \binom{q}{n+1}.
\end{aligned}$$

q.e.d.

Before getting to the main result of this section, one more technical lemma is needed.

LEMMA 2.15

$$\int_{M<r>} \log \frac{1}{|W| \mathbf{Z}^{\frac{n(n+1)}{2}}} \sigma - \int_{M<s>} \log \frac{1}{|W| \mathbf{Z}^{\frac{n(n+1)}{2}}} \sigma =$$

$$\frac{n(n+1)}{2} Ric_{\tau}(r, s) + \sum_{j=0}^{n} N_{f, a_j}(r, s) - N_{F_n}(r, s). \tag{2.32}$$

PROOF: By 1.6,

$$\begin{aligned}
Ric_{\tau}(r, s) &= N_{\Theta}(r, s) - \int_{M<r>} \log \mathbf{Z} \sigma + \int_{M<s>} \log \mathbf{Z} \sigma; \\
\frac{-n(n+1)}{2} Ric_{\tau}(r, s) &= \frac{-n(n+1)}{2} N_{\Theta}(r, s) + \int_{M<r>} \log \mathbf{Z}^{\frac{n(n+1)}{2}} \sigma - \\
&\quad \int_{M<s>} \log \mathbf{Z}^{\frac{n(n+1)}{2}} \sigma; \\
\frac{n(n+1)}{2} N_{\Theta}(r, s) &= \frac{n(n+1)}{2} Ric_{\tau}(r, s) + \int_{M<r>} \log \mathbf{Z}^{\frac{n(n+1)}{2}} \sigma - \\
&\quad \int_{M<s>} \log \mathbf{Z}^{\frac{n(n+1)}{2}} \sigma.
\end{aligned} \tag{2.33}$$

Also, by the Jensen formula and 2.27,

$$\begin{aligned}
-N_W(r, s) &= \int_{M<r>} \log \frac{1}{|W|} \sigma - \int_{M<s>} \log \frac{1}{|W|} \sigma \\
&= -N_{F_n}(r, s) + \frac{n(n+1)}{2} N_{\Theta}(r, s) + \sum_{j=0}^{n} N_{f, a_j}(r, s); \\
\frac{n(n+1)}{2} N_{\Theta}(r, s) &= N_{F_n}(r, s) - \sum_{j=0}^{n} N_{f, a_j}(r, s) + \int_{M<r>} \log \frac{1}{|W|} \sigma - \\
&\quad \int_{M<s>} \log \frac{1}{|W|} \sigma.
\end{aligned} \tag{2.34}$$

Setting 2.33 and 2.34 equal and combining the integrals,

$$\int_{M<r>} \log \frac{1}{|W|\mathbf{Z}^{\frac{n(n+1)}{2}}}\sigma - \int_{M<s>} \log \frac{1}{|W|\mathbf{Z}^{\frac{n(n+1)}{2}}}\sigma = \frac{n(n+1)}{2} Ric_\tau(r,s) + \sum_{j=0}^{n} N_{f,a_j}(r,s) - N_{F_n}(r,s).$$

q.e.d.

This then leads to the main result of this section.

THEOREM 2.16 [SECOND MAIN THEOREM FOR FIXED TARGETS] *Assume that* [A1]-[A11] *hold. Let* F_n *be the canonical representation section of order* n. *Then, for all* $0 < s < r$ *except for a set of finite measure,*

$$\begin{aligned}(q-n-1)T_f(r,s) + N_{F_n}(r,s) \ \underset{..}{\leq}\ & \sum_{g\in G} N_{f,g}(r,s) + \frac{n(n+1)}{2} Ric_\tau(r,s) + \\ & \binom{q}{n+1}\frac{1}{2}n^2(n+1)(1+\epsilon)\varsigma[7\log^+ T_f(r,s) + \\ & \log^+ Y_{\mathbb{B}}(r) + 7\log^+ r] + \\ & \binom{q}{n+1}\frac{7}{2}n^2(n-1)(1+\epsilon)\varsigma[\log^+ Ric_\tau(r,s) + \\ & \log^+ E(r)]. \end{aligned} \tag{2.35}$$

PROOF: By integrating the Pointwise Second Main Theorem over $M<r>$,

$$\begin{aligned}\int_{M<r>} \sum_{g\in G} \log \frac{1}{⟦f,g⟧}\sigma &= \sum_{g\in G}\int_{M<r>} \log\frac{1}{⟦f,g⟧}\sigma \\ &\leq \sum_{j=0}^{n}\int_{M<r>} \log \frac{1}{⟦f,a_j⟧}\sigma + \int_{M<r>} \log\frac{1}{|W|\mathbf{Z}^{\frac{n(n+1)}{2}}}\sigma \\ & \quad \sum_{B\in\mathfrak{B}}\int_{M<r>} \log^+ |W_B|\mathbf{Z}^{\frac{n(n+1)}{2}}\sigma + \\ & \quad \int_{M<r>} (q-n)\log\frac{n+1}{\Gamma(G)}\sigma + \varsigma\binom{q}{n+1} \\ &= \sum_{j=0}^{n} m_{f,a_j}(r) + \int_{M<r>} \log\frac{1}{|W|\mathbf{Z}^{\frac{n(n+1)}{2}}}\sigma + \\ & \quad \sum_{B\in\mathfrak{B}}\int_{M<r>} \log^+ |W_B|\mathbf{Z}^{\frac{n(n+1)}{2}}\sigma + \varsigma(q-n)\log\frac{n+1}{\Gamma(G)} + \\ & \quad \varsigma\binom{q}{n+1} \\ &= \sum_{j=0}^{n} m_{f,a_j}(r) + \left[\int_{M<s>} \log\frac{1}{|W|\mathbf{Z}^{\frac{n(n+1)}{2}}}\sigma + \right. \\ & \quad \left. \frac{n(n+1)}{2}Ric_\tau(r,s) + \sum_{j=0}^{n} N_{f,a_j}(r,s) - N_{F_n}(r,s)\right] + \end{aligned}$$

$$\sum_{B\in\mathfrak{B}}\int_{M<r>}\log^+|W_B|\mathbf{Z}^{\frac{n(n+1)}{2}}\sigma +$$
$$\varsigma(q-n)\log\frac{n+1}{\Gamma(G)}+\varsigma\binom{q}{n+1}$$

by 2.32. Then, through an application of 2.28,

$$\begin{aligned}
\sum_{g\in G}\int_{M<r>}\log\frac{1}{\|f,g\|}\sigma \;\underset{..}{\leq}\; & \sum_{j=0}^{n} m_{f,a_j}(r)+\sum_{j=0}^{n} N_{f,a_j}(r,s)+\frac{n(n+1)}{2}Ric_\tau(r,s)- \\
& N_{F_n}(r,s)+\int_{M<s>}\log\frac{1}{|W|\mathbf{Z}^{\frac{n(n+1)}{2}}}\sigma + \\
& \binom{q}{n+1}\{\,\frac{1}{2}n^2(n+1)(1+\frac{\epsilon}{2})\varsigma[7\log^+ T_f(r,s)+ \\
& \log^+ Y_{\mathbb{B}}(r)+7\log^+ r]+ \\
& \frac{7}{2}n^2(n-1)(1+\frac{\epsilon}{2})\varsigma[\log^+ Ric_\tau(r,s)+\log^+ E(r)]\,\}+ \\
& \varsigma(q-n)\log\frac{n+1}{\Gamma(G)}+\varsigma\binom{q}{n+1} \\
\underset{..}{\leq}\; & (n+1)T_f(r,s)+\frac{n(n+1)}{2}Ric_\tau(r,s)-N_{F_n}(r,s)+ \\
& \binom{q}{n+1}\frac{1}{2}n^2(n+1)(1+\epsilon)\varsigma[7\log^+ T_f(r,s)+ \\
& \log^+ Y_{\mathbb{B}}(r)+7\log^+ r]+ \\
& \binom{q}{n+1}\frac{7}{2}n^2(n-1)(1+\epsilon)\varsigma[\log^+ Ric_\tau(r,s)+ \\
& \log^+ E(r)],
\end{aligned}$$

by an application of the First Main Theorem.

Thus, by the First Main Theorem and since $m_{f,g}(r)=\int_{M<r>}\log\frac{1}{\|f,g\|}\sigma$,

$$\begin{aligned}
\sum_{g\in G}T_f(r,s)=qT_f(r,s) \;\underset{..}{\leq}\; & \sum_{g\in G}m_{f,g}(r)+\sum_{g\in G}N_{f,g}(r,s)-\sum_{g\in G}m_{f,g}(s) \\
\underset{..}{\leq}\; & \sum_{g\in G}N_{f,g}(r,s)-\sum_{g\in G}m_{f,g}(s)+(n+1)T_f(r,s)+ \\
& \frac{n(n+1)}{2}Ric_\tau(r,s)-N_{F_n}(r,s)+ \\
& \binom{q}{n+1}\frac{1}{2}n^2(n+1)(1+\epsilon)\varsigma[7\log^+ T_f(r,s)+ \\
& \log^+ Y_{\mathbb{B}}(r)+7\log^+ r]+ \\
& \binom{q}{n+1}\frac{7}{2}n^2(n-1)(1+\epsilon)\varsigma[\log^+ Ric_\tau(r,s)+ \\
& \log^+ E(r)].
\end{aligned}$$

Equivalently,

$$(q-n-1)T_f(r,s)+N_{F_n}(r,s)\;\underset{..}{\leq}\;\sum_{g\in G}N_{f,g}(r,s)+\frac{n(n+1)}{2}Ric_\tau(r,s)+$$

$$\binom{q}{n+1}\frac{1}{2}n^2(n+1)(1+\epsilon)\varsigma[7\log^+ T_f(r,s) +$$
$$\log^+ Y_{\mathbb{B}}(r) + 7\log^+ r] +$$
$$\binom{q}{n+1}\frac{7}{2}n^2(n-1)(1+\epsilon)\varsigma[\log^+ Ric_\tau(r,s) +$$
$$\log^+ E(r)].$$

q.e.d.

Chapter 3

3.1 Fields and vector spaces of meromorphic maps

To find a defect relation for slowly moving targets, Steinmetz [8] utilized a new idea for the Wronski determinant. Stoll [14] and Ru-Stoll [7] then extended this work to find the defect relation for holomorphic curves on $\mathbb{C}$. Bardis [2] then localized the techniques found in [14] and [7] to handle meromorphic maps on parabolic, covering manifolds. Now, these methods will be further extended to find the defect relation for meromorphic maps on parabolic manifolds. Some introductory definitions and results must first be considered.

Let V be a complex vector space. A subset G of $\mathbb{P}(V^*)$ is said to be in general position if each subset H of G containing $p+1$ elements, $p+1 \leq n+1$, spans a p-dimensional projective linear subspace of $\mathbb{P}(V^*)$. See Ru-Stoll [6], pp. 299-300. For equivalent characterizations of this concept, consider assumption (A11) in Chapter 2 and see Bardis [2], pp. 62, 87-88.

Next, the notion of general position must be extended to a finite set $\mathfrak{G}$ of meromorphic maps $g : M \rightarrow \mathbb{P}(V^*)$ where M is an m-dimensional connected, complex manifold. First of all, for every g in $\mathfrak{G}$, denote the indeterminacy of g by $I(g)$. Then, define the indeterminacy of $\mathfrak{G}$ by

$$I(\mathfrak{G}) = \bigcup_{g \in \mathfrak{G}} I(g). \tag{3.1}$$

Here, $I(\mathfrak{G})$ is analytic in M such that the dimension of $I(\mathfrak{G})$ is at most $m-2$. For all z in $M - I(\mathfrak{G})$, define the orbit $G(z) = \{g(z) | g \in \mathfrak{G}\}$. Furthermore, the orbit $\mathfrak{G}(z)$ is called faithful if $\#\mathfrak{G}(z) = \#\mathfrak{G}$. Then, $\mathfrak{G}$ is said to be in general position if there exists a point z_0 in $M - I(\mathfrak{G})$ such that its orbit $\mathfrak{G}(z_0)$ is in general position and faithful.

Now, denote by $\mathfrak{M}$ the field consisting of all meromorphic functions on M. Suppose $\mathfrak{K}$ is a subfield of $\mathfrak{M}$ such that $\mathbb{C}$ is a subset of $\mathfrak{K}$. Then, a meromorphic map $f : M \rightarrow \mathbb{P}(V)$ is called dependent on $\mathfrak{K}$ if $f_{\mathfrak{a}\mathfrak{b}}$ is in $\mathfrak{K}$ for all $\mathfrak{a}$ in V^*, $\mathfrak{b}$ in V_*^* such that (f, b) is free, where $b = \mathbb{P}(\mathfrak{b})$. Similarly, a meromorphic map $g : M \rightarrow \mathbb{P}(V^*)$ is called dependent on $\mathfrak{K}$ if $g_{\mathfrak{c}\mathfrak{d}}$ is in $\mathfrak{K}$ for all $\mathfrak{c}$ in V and $\mathfrak{d}$ in V_* such that (d, g) is free, where $d = \mathbb{P}(\mathfrak{d})$. Then, f is called linearly non-degenerated over $\mathfrak{K}$ if (f, g) is free for each meromorphic map $g : M \rightarrow \mathbb{P}(V^*)$ dependent on $\mathfrak{K}$.

As before, let $\mathfrak{G}$ denote a finite set of meromorphic maps on M such that $\#\mathfrak{G} \geq n+1$. Denote by $\mathfrak{K}_{\mathfrak{G}}$ the subfield of $\mathfrak{M}$ generated by all coordinate functions of all maps g in $\mathfrak{G}$. Note that $\mathbb{C}$ is a subset of $\mathfrak{K}_{\mathfrak{G}}$. Now, suppose that $\mathfrak{e} = (\mathfrak{e}_0, ..., \mathfrak{e}_n)$

is a base of V. Then, $\mathfrak{e}$ is called perfect for $\mathfrak{G}$ if and only if the following property holds:

> Take any subset P of $\mathfrak{G}$ with $\#P = n+1$. Take any enumeration $P = \{g_0, g_1, ..., g_n\}$. Take any representation $\mathfrak{w}_j : U \to V^*$ of g_j for $j = 0, 1, ..., n$ on a common, nonempty, open, connected subset U of M. Then, the following holds:
> $$< \mathfrak{w}_0 \wedge \mathfrak{w}_1 \wedge ... \wedge \mathfrak{w}_j, \mathfrak{e}_0 \wedge \mathfrak{e}_1 \wedge ... \wedge \mathfrak{e}_j > \not\equiv 0$$
> for $j = 0, 1, ..., n$.

It is sufficient that this property holds for one selection U, $\mathfrak{w}_0, ..., \mathfrak{w}_n$ only. If $\mathfrak{G}$ is in general position, then there exists an orthonormal, perfect base of V for $\mathfrak{G}$.

For the remainder of this chapter, the following assumptions will be made:

(B1) M is a connected complex manifold of dimension m.

(B2) V is a hermitian vector space of dimension $n+1 > 1$.

(B3) $\mathfrak{G}$ is a finite set of meromorphic maps $g : M \to \mathbb{P}(V^*)$ such that $\#\mathfrak{G} \geq n+1$.

(B4) $\mathfrak{G}$ is in general position.

(B5) $\mathfrak{e} = (\mathfrak{e}_0, ..., \mathfrak{e}_n)$ is an orthonormal, perfect base of V for $\mathfrak{G}$. Let $\mathfrak{a} = (\mathfrak{a}_0, ..., \mathfrak{a}_n)$ be the dual base of $\mathfrak{e}$. Define $e_j = \mathbb{P}(\mathfrak{e}_j)$ and $a_j = \mathbb{P}(\mathfrak{a}_j)$ for $j = 0, 1, ..., n$.

(B6) Let $f : M \to \mathbb{P}(V)$ be a meromorphic map which is linearly non-degenerated over $\mathfrak{K}_{\mathfrak{G}}$.

(B7) Let $\mathfrak{U} = \{U_\lambda\}_{\lambda \in \Lambda}$ be a covering of M. Here, each U_λ is a non-empty, open, connected Stein subset of M such that $H^2(U_\lambda, \mathbb{Z}) = 0$.

Under these assumptions, a considerable number of various objects will be constructed. The standard notation for a covering will be retained. For p in $\mathbb{Z}_+$ and $\lambda = (\lambda_0, ..., \lambda_p)$ in Λ^{p+1},

$$\begin{aligned} U_\lambda &= U_{\lambda_0,...,\lambda_p} = U_{\lambda_0} \cap ... \cap U_{\lambda_p}; \\ \Lambda[p] &= \{\lambda \in \Lambda^{p+1} | U_\lambda \neq \emptyset\}. \end{aligned}$$

By (B7), every meromorphic map from M into a complex projective space has a reduced representation on each U_λ. Also, every holomorphic line bundle on M is trivial over each U_λ.

For every λ in Λ, select a reduced representation $\mathfrak{v}_\lambda$ of f on U_λ given by

$$\mathfrak{v}_\lambda = \sum_{i=0}^{n} v_\lambda^i \mathfrak{e}_i : U_\lambda \to V$$

where $v_\lambda^i =< \mathfrak{v}_\lambda, \mathfrak{a}_i >$. By (B6), $v_\lambda^i \not\equiv 0$ for each $j = 0, ..., n$ and each λ in Λ. Then, for every (λ, μ) in $\Lambda[1]$, there exist unique holomorphic transition functions $v_{\lambda\mu} : U_\lambda \cap U_\mu \to \mathbb{C}_*$ satisfying

$$\mathfrak{v}_\lambda = v_{\lambda\mu} \mathfrak{v}_\mu, \tag{3.2}$$

$$v_\lambda^i = v_{\lambda\mu} v_\mu^i \tag{3.3}$$

for $i = 0, ..., n$. These transition functions are such that for λ in Λ, (λ, μ) in $\Lambda[1]$, and (λ, μ, ρ) in $\Lambda[2]$,

$$\begin{array}{lrl} i. & v_{\lambda\lambda} = 1 & \text{on } U_\lambda; \\ ii. & v_{\lambda\mu}v_{\mu\lambda} = 1 & \text{on } U_\lambda \cap U_\mu; \\ iii. & v_{\lambda\mu}v_{\mu\rho}v_{\rho\lambda} = 1 & \text{on } U_\lambda \cap U_\mu \cap U_\rho. \end{array}$$

Thus, $\{v_{\lambda\mu}\}_{(\lambda,\mu)\in\Lambda[1]}$ forms a basic cocycle of L_f, the hyperplane section bundle of f.

If $\overset{\triangle}{\mathfrak{v}}_\lambda$ is the holomorphic frame of L_f over U_λ associated to $\mathfrak{v}_\lambda$, then on $U_\lambda \cap U_\mu$,

$$\overset{\triangle}{\mathfrak{v}}_\mu \quad = \quad v_{\lambda\mu}\, \overset{\triangle}{\mathfrak{v}}_\lambda\, .$$

Also, suppose $\tilde{\mathfrak{v}}_\lambda$ is the holomorphic section of $M \times V = V_M$ over U_λ associated to $\mathfrak{v}_\lambda$ given by $\tilde{\mathfrak{v}}_\lambda(x) = (x, \mathfrak{v}_\lambda(x))$ for all x in U_λ. Then, the reduced representation section F_f of $V_M \otimes L_f$ associated to f is such that

$$F_f|U_\lambda \quad = \quad \tilde{\mathfrak{v}}_\lambda \otimes \overset{\triangle}{\mathfrak{v}}_\lambda\, .$$

Denote $f_i = f_{\mathfrak{a}_i\mathfrak{a}_0}$. Therefore, $f_i|U_\lambda = \frac{v_\lambda^i}{v_\lambda^0}$.

Now, choose any g in $\mathfrak{G}$. For every λ in Λ, select a reduced representation $\mathfrak{w}_\lambda$ of g on U_λ given by

$$\mathfrak{w}_\lambda(g) \quad = \quad \sum_{i=0}^{n} w_\lambda^i(g)\mathfrak{a}_i : U_\lambda \to V^*$$

where $w_\lambda^i(g) =< \mathfrak{e}_i, \mathfrak{w}_\lambda(g) >$. Then, for every (λ, μ) in $\Lambda[1]$, there exist unique holomorphic transition functions $w_{\lambda\mu}(g) : U_\lambda \cap U_\mu \to \mathbb{C}_*$ satisfying

$$\mathfrak{w}_\lambda(g) = w_{\lambda\mu}(g)\mathfrak{w}_\mu(g) \quad \text{and} \quad w_\lambda^i(g) = w_{\lambda\mu}(g)w_\mu^i(g)$$

for $i = 0, ..., n$. These transition functions are such that for λ in Λ, (λ, μ) in $\Lambda[1]$, and (λ, μ, ρ) in $\Lambda[2]$,

$$\begin{array}{lrl} i. & w_{\lambda\lambda}(g) = 1 & \text{on } U_\lambda; \\ ii. & w_{\lambda\mu}(g)w_{\mu\lambda}(g) = 1 & \text{on } U_\lambda \cap U_\mu; \\ iii. & w_{\lambda\mu}(g)w_{\mu\rho}(g)w_{\rho\lambda}(g) = 1 & \text{on } U_\lambda \cap U_\mu \cap U_\rho. \end{array}$$

Hence, $\{w_{\lambda\mu}(g)\}_{(\lambda,\mu)\in\Lambda[1]}$ forms a basic cocycle of L_g, the hyperplane section bundle of g. If $\overset{\triangle}{\mathfrak{w}}_\lambda(g)$ is the holomorphic frame of L_g over U_λ associated to $\mathfrak{w}_\lambda(g)$, then on $U_\lambda \cap U_\mu$,

$$\overset{\triangle}{\mathfrak{w}}_\mu(g) \quad = \quad w_{\lambda\mu}(g)\, \overset{\triangle}{\mathfrak{w}}_\lambda(g).$$

In addition, the reduced representation section F_g of $V_M^* \otimes L_g$ associated to g is such that

$$F_g|U_\lambda \quad = \quad \tilde{\mathfrak{w}}_\lambda(g) \otimes \overset{\triangle}{\mathfrak{w}_\lambda}(g).$$

If $g(i) = g_{\mathfrak{e}_i \mathfrak{e}_0}$, then $g(i)|U_\lambda = \frac{w_\lambda^i(g)}{w_\lambda^0(g)}$. Also, write $\mathfrak{g}(g) = \sum_{i=0}^n g(i)\mathfrak{a}_i$ such that $\mathfrak{g}(g)|U_\lambda = \frac{\mathfrak{w}_\lambda(g)}{w_\lambda^0(g)}$.

Let Ψ be a nonempty set of meromorphic functions on M. Then, (T, Δ) is called common (or universal) denominator bundle for Ψ if an only if the following conditions hold:

1. T is a holomorphic line bundle over M.
2. $\Delta \not\equiv 0$ is a holomorphic section of T.
3. $\phi\Delta$ is a holomorphic section of T for all ϕ in Ψ.
4. There exists an analytic set E in M with dimension at most $m - 2$ such that for every z in $M - E$, there exists ϕ in Ψ with $(\phi\Delta)(z)$ nonzero.

The pair (T, Δ) is unique up to a natural isomorphism according to Lemma 3.1.1 of Bardis [2]. The section Δ is called a common (or universal) denominator for Ψ. If Ψ is finite and 1 is in Ψ, then there exists a common denominator bundle for Ψ. See Theorem 3.1.3 of Bardis [2].

Based on (B1) through (B7), the construction will now be continued. Define

$$\Phi \quad = \quad \{g_{\mathfrak{e}_j \mathfrak{e}_0} | g \in \mathfrak{G} \text{ and } j = 0, ..., n\}.$$

Then Φ is finite with 1 in Φ. Let (T, Δ) be the common denominator bundle for Φ. Suppose $\mathfrak{L}$ is the complex vector space generated by Φ. Note that $\mathfrak{L}$ is contained in $\mathfrak{K}_{\mathfrak{G}}$. Also, (T, Δ) is the common denominator bundle for $\mathfrak{L}$. Denote the dual vector space by $Y = \mathfrak{L}^*$.

Since $T|U_\lambda$ is trivial for each λ in Λ, then there exists a holomorphic frame atlas $\{u_\lambda\}_{\lambda \in \Lambda}$ of T over $\mathfrak{U}$. Thus, u_λ is a holomorphic frame of T over U_λ for each λ in Λ. For (λ, μ) in $\Lambda[1]$, there exists a nowhere zero holomorphic function $T_{\lambda\mu} : U_{\lambda\mu} \to \mathbb{C}_*$ such that on $U_{\lambda\mu}$,

$$u_\lambda \quad = \quad T_{\mu\lambda} u_\mu.$$

Note that $T_{\lambda\lambda} = 1, T_{\lambda\mu}T_{\mu\lambda} = 1$ on $U_{\lambda\mu}$, and $T_{\mu\lambda}T_{\lambda\rho}T_{\rho\mu} = 1$ on $U_{\lambda\mu\rho}$ for (λ, μ, ρ) in $\Lambda[2]$. Also, a unique holomorphic function $\Delta_\lambda \not\equiv 0$ exists on U_λ such that

$$\Delta|U_\lambda \quad = \quad \Delta_\lambda u_\lambda.$$

Then, on $U_{\lambda\mu}$ for (λ, μ) in $\Lambda[1]$,

$$\Delta_\lambda \quad = \quad T_{\lambda\mu}\Delta_\mu. \tag{3.4}$$

For λ in Λ, a map $\mathfrak{y}_\lambda : U_\lambda \to Y$ will now be introduced. For z in U_λ, $\mathfrak{y}_\lambda(z)$ in $Y = \mathfrak{L}(\Phi)^*$ must be defined. For any ϕ in $\mathfrak{L}(\Phi)$, $\phi\Delta = \phi\Delta_\lambda u_\lambda$ is holomorphic on U_λ and thus $\phi\Delta_\lambda$ is holomorphic on U_λ. Then, define

$$< \phi, \mathfrak{y}_\lambda(z) > \;=\; \mathfrak{y}_\lambda(z)(\phi) = (\phi\Delta_\lambda)(z).$$

Here, $\mathfrak{y}_\lambda : U_\lambda \to Y$ is holomorphic, $\mathfrak{y}_\lambda(z) \neq 0$ if z in $U_\lambda - E$, and on $U_{\lambda\mu}$,

$$\mathfrak{y}_\lambda \;=\; T_{\lambda\mu}\mathfrak{y}_\mu \tag{3.5}$$

for (λ, μ) in $\Lambda[1]$. See also pp. 75-76 of Bardis [2]. Thus, on $U_{\lambda\mu}$ for (λ,μ) in $\Lambda[1]$,

$$\begin{aligned} \mathfrak{y}_\lambda \otimes u_\lambda &= T_{\lambda\mu}\mathfrak{y}_\mu \otimes u_\lambda \\ &= \mathfrak{y}_\mu \otimes u_\mu. \end{aligned}$$

Consequently, there is a unique holomorphic section $\mathfrak{y}$ of $Y_M \otimes T$ over M given by

$$\mathfrak{y}|U_\lambda \;=\; \mathfrak{y}_\lambda \otimes u_\lambda.$$

Furthermore, on U_λ, $< \phi, \mathfrak{y}(z) >=< \phi, \mathfrak{y}_\lambda(z) > u_\lambda(z) = (\phi\Delta_\lambda)(z)u_\lambda(z) = (\phi\Delta)(z)$. Therefore, for each ϕ in $\mathfrak{L}(\Phi)$ and z in U_λ,

$$< \phi, \mathfrak{y}(z) > \;=\; (\phi\Delta)(z).$$

Also, the adjunct map to $\mathfrak{G}$ is defined to be the meromorphic map $y : M \to \mathbb{P}(Y)$ given by

$$y(z) \;=\; \mathbb{P}(\mathfrak{y}_\lambda(z))$$

for λ in Λ and z in $U_\lambda - E$.

Select p in $\mathbb{N}$. Denote by $\mathfrak{L}_p$ the complex vector space generated by

$$\Phi_p \;=\; \{\phi_1 \cdots \phi_p | \phi_j \in \Phi, j = 1, ..., p\}.$$

Note that $\mathfrak{L}_p$ is contained in $\mathfrak{K}_{\mathfrak{G}}$. Denote by $\bigodot_p Y$ the p-fold symmetric tensor product of Y. Write $\xi_1 \odot \xi_2 \odot \cdots \odot \xi_p$ for the symmetric tensor product of ξ_j in Y for $j = 1, 2, ..., p$. Additionally, for $\xi = \xi_1 = \xi_2 = ... = \xi_p$, write $\xi^p = \xi_1 \odot \xi_2 \odot \cdots \odot \xi_p$. For x_j in $\mathbb{P}(Y)$, then $x_j = \mathbb{P}(\xi_j)$ and $\xi_1 \odot \xi_2 \odot \cdots \odot \xi_p \not\equiv 0$, which implies that $x_1 \odot x_2 \odot \cdots \odot x_p = \mathbb{P}(\xi_1 \odot \xi_2 \odot \cdots \odot \xi_p)$. Thus, $x = \mathbb{P}(\xi)$ implies $x^p = \mathbb{P}(\xi^p)$. Therefore, $\mathfrak{y}_\lambda^p : U_\lambda \to \bigodot_p Y$ and $y^p : M \to \mathbb{P}(\bigodot_p Y)$ are well-defined. Next, $\mathfrak{y}^p$ is a holomorphic section of $(\bigodot_p Y) \otimes T^p$ given by

$$\mathfrak{y}^p|U_\lambda \;=\; \mathfrak{y}_\lambda^p \otimes u_\lambda^p$$

for all λ in Λ. Let $Y(p)$ be the smallest linear subspace of $\bigodot_p Y$ satisfying

$$y^p(M) \subseteq \mathbb{P}(Y(p)),$$

which is equivalent to $Y(p)$ as the smallest linear subspace of $\bigodot_p Y$ satisfying

$$\mathfrak{y}_\lambda^p(U_\lambda) \subseteq Y(p)$$

for all λ in Λ. Denote the dimension of $Y(p)$ as $q(p)$. See Lemma 3.1.8 of Bardis [2] for more details.

Suppose $\iota : Y(p) \to \underset{p}{\odot} Y$ is the inclusion map. There exists a unique surjective linear map $\sigma : \underset{p}{\odot} \mathfrak{L} \to \mathfrak{L}_p$ given by

$$\sigma(\phi_1 \odot \odot \odot \phi_p) \quad = \quad \phi_1 \cdots \phi_p,$$

for ϕ_j in $\mathfrak{L}$ for $j = 1, ..., p$. See Lemma 3.6 of Stoll [14]. Then, the dual map $\sigma^* : \mathfrak{L}_p^* \to \underset{p}{\odot} Y$ is injective. In fact, $\sigma^*(\mathfrak{L}_p^*) = Y(p)$ and an isomorphism is given by $\rho = \sigma^* : \mathfrak{L}_p^* \to Y(p)$. See Lemma 3.1.13 of Bardis [2]. Therefore, $\sigma^* = \iota \circ \rho$ and $\sigma = \rho^* \circ \iota^* : \underset{p}{\otimes} \mathfrak{L} \to \mathfrak{L}_p$.

Choose p, s in $\mathbb{N}$. Take g in $\mathfrak{L}_s$ and $\tilde{g}$ in $\underset{s}{\odot} \mathfrak{L}$ such that $\sigma(\tilde{g}) = g \not\equiv 0$. In turn, define linear maps

$$\dot{H}_{\tilde{g}} \;:\; \underset{p}{\odot} \mathfrak{L} \to \underset{p+s}{\odot} \mathfrak{L},$$
$$H_g \;:\; \mathfrak{L}_p \to \mathfrak{L}_{p+s}$$

by $\dot{H}_{\tilde{g}}(\psi) = \tilde{g} \odot \psi$ for every ψ in $\underset{p}{\odot} \mathfrak{L}$ and $H_g(\psi) = g \cdot \psi$ for every ψ in $\mathfrak{L}_p$. If $\sigma(\tilde{g}) = g$, then $\sigma \circ \dot{H}_{\tilde{g}} = H_g \circ \sigma$. If $\tilde{g} \equiv 0 \equiv g$, then $\dot{H}_{\tilde{g}} \equiv 0 \equiv H_g$, respectively. Also, if $\tilde{g} \not\equiv 0 \not\equiv g$, then $\dot{H}_{\tilde{g}}$ and H_g are injective, respectively.

The following dual diagram commutes:

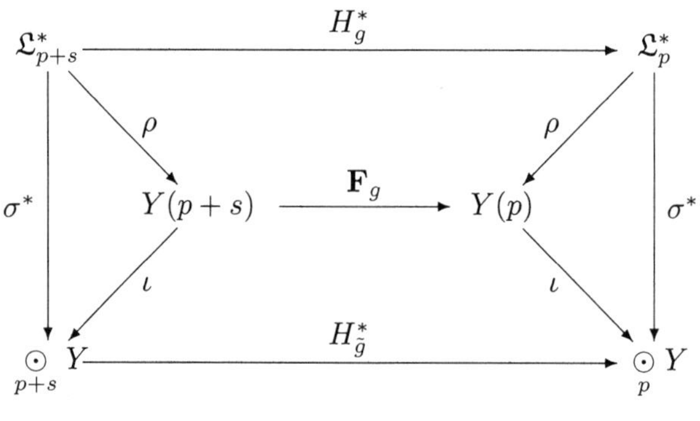

The map $\mathbf{F}_g = \mathbf{F}_g[s,p] = \rho \circ H_g^* \circ \rho^{-1} : Y(p+s) \to Y(p)$ is surjective and linear . Now, suppose g is in $\mathfrak{L}_s$. On U_λ for λ in Λ,

$$\mathbf{F}_g[s,p] \circ \mathfrak{y}_\lambda^{p+s} \quad = \quad \Delta_\lambda^s g \mathfrak{y}_\lambda^p. \tag{3.6}$$

See Proposition 3.1.14 of Bardis [2]. Also, see pp. 110-111 of Ru-Stoll [7].

Take a hermitian metric on $\mathfrak{L}$ such that $\| 1 \| = 1$. This defines a hermitian metric on $Y = \mathfrak{L}^*$. In turn, this defines hermitian metrics on $\underset{p}{\odot} \mathfrak{L}$ and $\underset{p}{\odot} Y$ which then restricts to a hermitian metric on the linear subspace $Y(p)$. Since $\rho : \mathfrak{L}_p^* \to Y(p)$ is an isomorphism, there exists one and only one hermitian metric on $\mathfrak{L}_p^*$ such that ρ is an isometry. Then, the dual hermitian metric on $\mathfrak{L}_p$ is defined.

Let $X_g(s,p)$ be the kernel of $\mathbf{F}_g(s,p)$ and denote the inclusion map of $X_g(s,p)$ into $Y(p+s)$ by $\iota_g = \iota_g[s,p]$. The hermitian metric on $Y(p+s)$ will define a direct sum decomposition:

$$0 \rightleftarrows X_g(s,p) \underset{Q_g}{\overset{\iota_g}{\rightleftarrows}} Y(p+s) \underset{\eta_g}{\overset{\mathbf{F}_g}{\rightleftarrows}} Y(p) \rightleftarrows 0$$

$Y(p+s)$ can therefore be viewed as $X_g(s,p) \oplus X_g(s,p)^{\perp} = X_g(s,p) \oplus Y(p)$. In addition, $\iota_g = \iota_g[s,p]$ and $\eta_g = \eta_g[s,p]$ are injections; $Q_g = Q_g[s,p]$ and $\mathbf{F}_g = \mathbf{F}_g[s,p]$ are projections; and $Q_g \circ \iota_g$, $\mathbf{F}_g \circ \eta_g$, and $\iota_g \circ Q_g + \eta_g \circ \mathbf{F}_g$ are identities. For $g \equiv 1 \equiv 1^s$ in $\mathfrak{L}_s$, the index of 1 is dropped. There exists an isometric isomorphism

$$\beta_g = \beta_g[s,p] : X_g(s,p) \to X(s,p)$$

which is chosen at random for each $0 \not\equiv g$ in $\mathfrak{L}_s$ such that $\beta = \beta_1$ is the identity. See pp. 111-112 of Ru-Stoll [7] for more details.

The linear map given by

$$\tilde{\mathbf{F}}_g = \tilde{\mathbf{F}}_g[s,p] = \eta \circ \mathbf{F}_g + \iota \circ \beta_g \circ Q_g : Y(p+s) \to Y(p+s)$$

is an isomorphism. See Proposition 3.2 of Ru-Stoll [7]. Then, the following two exact sequences can be considered:

$$\begin{array}{ccccccccc}
0 & \rightleftarrows & X_g(s,p) & \underset{Q_g}{\overset{\iota_g}{\rightleftarrows}} & Y(p+s) & \underset{\eta_g}{\overset{\mathbf{F}_g}{\rightleftarrows}} & Y(p) & \rightleftarrows & 0 \\
 & & \downarrow \beta_g & & \downarrow \tilde{\mathbf{F}}_g & & \downarrow id & & \\
0 & \rightleftarrows & X(s,p) & \underset{Q}{\overset{\iota}{\rightleftarrows}} & Y(p+s) & \underset{\eta}{\overset{\mathbf{F}}{\rightleftarrows}} & Y(p) & \rightleftarrows & 0
\end{array}$$

For all λ in Λ, define a holomorphic map

$$\mathfrak{x}_{\lambda g} = \mathfrak{x}_{\lambda g}[s,p] = \iota \circ \beta_g \circ Q_g \circ \mathfrak{y}_\lambda^{p+s} : U_\lambda \to Y(p+s)$$

which satisfies

$$\tilde{\mathbf{F}}_g \circ \mathfrak{y}_\lambda^{p+s} = \Delta_\lambda^s g \mathfrak{y}_\lambda^p + \mathfrak{x}_{\lambda g} \tag{3.7}$$

on U_λ for all λ in Λ. Notice the difference between 3.6 and 3.7. See pp. 85-86 of Bardis [2] and pp. 112-113 of Ru-Stoll [7] for further detail.

3.2 Steinmetz map and automorphisms of $Z(p)$

Select p in $\mathbb{N}$ such that $p > n+1$. The vector space

$$Z(p) \quad = \quad Y(p) \oplus Y(p+1)^n$$

has dimension $k(p)+1 = q(p)+nq(p+1)$. For every λ in Λ, define the holomorphic vector function $\mathfrak{h}_{[p],\lambda} = \mathfrak{h}_\lambda : U_\lambda \to Z(p)$ by

$$\mathfrak{h}_\lambda \quad = \quad \Delta_\lambda v_\lambda^0 \mathfrak{y}_\lambda^p \oplus \bigoplus_{j=1}^{n} v_\lambda^j \mathfrak{y}_\lambda^{p+1}.$$

Then, the transition formulas 3.3, 3.4, and 3.5 can be used to derive the transition formula for $\mathfrak{h}_\lambda$. It can be shown that on $U_{\lambda\mu}$ for (λ,μ) in $\Lambda[1]$,

$$\mathfrak{h}_\lambda \quad = \quad T_{\lambda\mu}^{p+1} v_{\lambda\mu} \mathfrak{h}_\mu. \tag{3.8}$$

See Lemma 3.2.1 of Bardis [2] for the details. Then, there exists a unique meromorphic map $h_{[p]} = h : M \to \mathbb{P}(Z(p))$ with $h|U_\lambda = \mathbb{P} \circ \mathfrak{h}_\lambda$ for every λ in Λ. The map $h_{[p]}$ is known as the pth Steinmetz map for $(\mathfrak{G}, f)$.

Let $\mathfrak{T} = \{\alpha : \mathbb{Z}[0,n] \to \mathfrak{G} | \alpha \text{ is injective }\}$. Since $\mathfrak{G}$ has at least $n+1$ elements, $\mathfrak{T}$ is nonempty. Select any α in $\mathfrak{T}$ and p in $\mathbb{N}$ with $p > \frac{n(n+1)}{2}$. Make the following abbreviations as done on pp. 93-94 of Bardis [2]:

1. $g_j \quad = \quad \alpha(j) : M \to \mathbb{P}(V^*)$ for j in $\mathbb{Z}[0,n]$.
2. $\mathfrak{w}_\lambda^j \quad = \quad \mathfrak{w}_\lambda(g_j) : U_\lambda \to V^*$ for j in $\mathbb{Z}[0,n], \lambda$ in Λ.
3. $g_{ji} \quad = \quad g_j(i) : M \to \mathbb{C}$ for j in $\mathbb{Z}[0,n]$, i in $\mathbb{Z}[0,n]$.
4. $w_\lambda^{ji} \quad = \quad w_\lambda^i(g_j) : U_\lambda \to \mathbb{C}$ for j in $\mathbb{Z}[0,n]$, i in $\mathbb{Z}[0,n]$.
5. $\mathfrak{g}_j \quad = \quad \mathfrak{g}(g_j) : M \to V^*$ for j in $\mathbb{Z}[0,n]$.

Then,

$$\mathfrak{g}_j | U_\lambda \quad = \quad \frac{\mathfrak{w}_\lambda^j}{w_\lambda^{j0}} = \frac{\mathfrak{w}_\lambda(g_j)}{w_\lambda^0(g_j)}.$$

For j in $\mathbb{Z}[0,n]$, i in $\mathbb{Z}[0,n]$, and λ in Λ,

$$\mathfrak{g}_j \quad = \quad \sum_{j=0}^{n} g_{ji} \mathfrak{a}_i,$$

such that $g_{ji} = \frac{w_\lambda^{ji}}{w_\lambda^{j0}}$ and $g_{j0} = 1$.

Next, for $j = 0, ..., n$ and λ in Λ, the linear maps

$$M_0 : Z(p) \to Y(p),\ M_j : Z(p) \to Y(p+1),\ L_j : Z(p) \to Z(p)$$

and the holomorphic maps

$$\mathfrak{h}_{\lambda,-1} = \mathfrak{h}_\lambda : U_\lambda \to Z(p), \qquad \mathfrak{h}_{\lambda,j} : U_\lambda \to Z(p)$$

will be inductively defined so that for $j = 0, ..., n$ and λ in Λ,

$$L_j \circ \mathfrak{h}_{\lambda,j-1} = \mathfrak{h}_{\lambda,j}.$$

Furthermore, for $0 \leq j \leq n$, each L_j will be constructed to be an automorphism of $Z(p)$.

To begin the inductive process, select $\mathfrak{z} = \mathfrak{z}_0 \oplus ... \oplus \mathfrak{z}_n$ in $Z(p)$ with z_0 in $Y(p)$ and z_i in $Y(p+1)$ for $i = 1, ..., n$. Define $M_0 : Z(p) \to Y(p)$, $L_0 : Z(p) \to Z(p)$, and $\mathfrak{h}_{\lambda,0} : U_\lambda \to Z(p)$ by

$$\begin{aligned} M_0(\mathfrak{z}) &= z_0 + \sum_{i=1}^{n} \mathbf{F}_{g_{0,i}}[1,p](z_i), \\ L_0(\mathfrak{z}) &= M_0(\mathfrak{z}) \oplus \mathfrak{z}_1 \oplus ... \oplus \mathfrak{z}_n, \\ \mathfrak{h}_{\lambda,0} &= <\mathfrak{v}_\lambda, \mathfrak{g}_0> \Delta_\lambda \mathfrak{y}_\lambda^p \oplus \bigoplus_{i=1}^{n} v_\lambda^i \mathfrak{y}_\lambda^{p+1}. \end{aligned}$$

Abbreviate $e_{\lambda,0} =<\mathfrak{v}_\lambda, \mathfrak{g}_0> \Delta_\lambda \mathfrak{y}_\lambda^p$ and thus $\mathfrak{h}_{\lambda,0} = e_{\lambda,0} \oplus \bigoplus_{i=1}^{n} v_\lambda^i \mathfrak{y}_\lambda^{p+1}$. It can be shown (see Bardis [2], pp. 94-94) that $M_0 \circ \mathfrak{h}_{\lambda,-1} = M_0 \circ \mathfrak{h}_\lambda = e_{\lambda,0}$. Therefore, on U_λ for all λ in Λ,

$$\begin{aligned} L_0 \circ \mathfrak{h}_{\lambda,-1} &= e_{\lambda,0} \oplus \bigoplus_{i=1}^{n} v_\lambda^i \mathfrak{y}_\lambda^{p+1} \\ &= \mathfrak{h}_{\lambda,0}. \end{aligned}$$

Additionally, by p. 115 of Ru-Stoll [7], L_0 is an isomorphism.

Now, define $\gamma_{1i} =<\mathfrak{g}_0 \wedge \mathfrak{g}_1, \mathfrak{e}_0 \wedge \mathfrak{e}_i>= g_{1i} - g_{0i}$. Here, γ_{1i} is in $\mathfrak{L}$ for all $i = 0, ..., n$. Then, $g_{10} - g_{00} = 1 - 1 = 0$ and

$$\begin{vmatrix} 1 & g_{01} \\ 1 & g_{11} \end{vmatrix} \not\equiv 0$$

since $\mathfrak{e}_0, ..., \mathfrak{e}_n$ is a perfect base of V. Define $M_1 : Z(p) \to Y(p+1)$, $L_1 : Z(p) \to Z(p)$, and $\mathfrak{h}_{\lambda,1} : U_\lambda \to Z(p)$ by

$$\begin{aligned} M_1(z_0 \oplus ... \oplus z_n) &= z_0 + \tilde{\mathbf{F}}_{\gamma_{11}}[1,p](z_1) + \sum_{i=2}^{n} \mathbf{F}_{\gamma_{1,i}}[1,p](z_i); \\ L_1(z_0 \oplus ... \oplus z_n) &= z_0 \oplus M_1(z_0 \oplus ... \oplus z_n) \oplus z_2 \oplus ... \oplus z_n; \\ \mathfrak{h}_{\lambda,1} &= <\mathfrak{v}_\lambda, \mathfrak{g}_0> \Delta_\lambda \mathfrak{y}_\lambda^p \oplus (<\mathfrak{v}_\lambda, \mathfrak{g}_1> \Delta_\lambda \mathfrak{y}_\lambda^p + \\ &\quad v_\lambda^1 \mathfrak{x}_{\lambda\gamma_{11}}[1,p]) \oplus \bigoplus_{i=2}^{n} v_\lambda^i \mathfrak{y}_\lambda^{p+1}. \end{aligned}$$

Abbreviate $e_{\lambda,1} =<\mathfrak{v}_\lambda, \mathfrak{g}_1> \Delta_\lambda \mathfrak{y}_\lambda^p + v_\lambda^1 \mathfrak{x}_{\lambda\gamma_{11}}[1,p]$ and thus $\mathfrak{h}_{\lambda,1} = e_{\lambda,0} \oplus e_{\lambda,1} \oplus \bigoplus_{i=2}^{n} v_\lambda^i \mathfrak{y}_\lambda^{p+1}$. It follows (see Proposition 3.2.2 of Bardis [2]) that $M_1 \circ \mathfrak{h}_{\lambda,0} = e_{\lambda,1}$. Therefore, on U_λ for all λ in Λ,

$$\begin{aligned} L_1 \circ \mathfrak{h}_{\lambda,0} &= e_{\lambda,0} \oplus e_{\lambda,1} \oplus \bigoplus_{i=2}^{n} v_\lambda^i \mathfrak{y}_\lambda^{p+1} \\ &= \mathfrak{h}_{\lambda,1}. \end{aligned}$$

Additionally, by Proposition 4.2 of Ru-Stoll [7], L_1 is an isomorphism.

Continuing inductively, let $\gamma_{2i} =< \mathfrak{g}_0 \wedge \mathfrak{g}_1 \wedge \mathfrak{g}_2, \mathfrak{e}_0 \wedge \mathfrak{e}_1 \wedge \mathfrak{e}_i >$. Here, γ_{2i} is in $\mathfrak{L}_2$ for all $i = 0, ..., n$. Note that $\gamma_{20} \equiv \gamma_{21} \equiv 0 \not\equiv \gamma_{22}$. Next, define $h_0 = < \mathfrak{g}_1 \wedge \mathfrak{g}_2, \mathfrak{e}_0 \wedge \mathfrak{e}_1 >$, $h_1 =< \mathfrak{g}_2 \wedge \mathfrak{g}_0, \mathfrak{e}_0 \wedge \mathfrak{e}_1 >$, and $h_2 =< \mathfrak{g}_0 \wedge \mathfrak{g}_1, \mathfrak{e}_0 \wedge \mathfrak{e}_1 >= \gamma_{11}$. Then, for $0 \leq k \leq 2$, each h_k is in $\mathfrak{L}$ and

$$\gamma_{2i} = g_{0i}h_0 + g_{1i}h_1 + g_{2i}h_2.$$

Also, define the two linear maps $M_2 : Z(p) \to Y(p+1)$ and $L_2 : Z(p) \to Z(p)$ and the holomorphic map $\mathfrak{h}_{\lambda,2} : U_\lambda \to Z(p)$ by

$$\begin{aligned} M_2(z_0 \oplus ... \oplus z_n) &= -\mathbf{F}_{h_0}[1, p-1](z_0) - \mathbf{F}_{h_1}[1, p-1](\mathbf{F}[1, p](z_1)) + \\ &\quad \tilde{\mathbf{F}}_{\gamma_{22}}[2, p-1](z_2) + \sum_{i=3}^{n} \mathbf{F}_{\gamma_{2,i}}[2, p-1](z_i); \\ L_2(z_0 \oplus ... \oplus z_n) &= z_0 \oplus z_1 \oplus M_2(z_0 \oplus ... \oplus z_n) \oplus z_3 \oplus ... \oplus z_n; \\ \mathfrak{h}_{\lambda,2} &= < \mathfrak{v}_\lambda, \mathfrak{g}_0 > \Delta_\lambda \mathfrak{y}_\lambda^p \oplus (< \mathfrak{v}_\lambda, \mathfrak{g}_1 > \Delta_\lambda \mathfrak{y}_\lambda^p + v_\lambda^1 \mathfrak{x}_{\lambda\gamma_{11}}[1, p]) \oplus \\ &\quad (< \mathfrak{v}_\lambda, \mathfrak{g}_2 > \gamma_{11} \Delta_\lambda^2 \mathfrak{y}_\lambda^{p-1} + v_\lambda^2 \mathfrak{x}_{\lambda\gamma_{22}}[2, p-1]) \oplus \bigoplus_{i=3}^{n} v_\lambda^i \mathfrak{y}_\lambda^{p+1}. \end{aligned}$$

Abbreviate $e_{\lambda,2} =< \mathfrak{v}_\lambda, \mathfrak{g}_2 > \gamma_{11}\Delta_\lambda^2 \mathfrak{y}_\lambda^{p-1} + v_\lambda^2 \mathfrak{x}_{\lambda\gamma_{22}}[2, p-1]$ and consequently $\mathfrak{h}_{\lambda,2} = e_{\lambda,0} \oplus e_{\lambda,1} \oplus e_{\lambda,2} \oplus \bigoplus_{i=3}^{n} v_\lambda^i \mathfrak{y}_\lambda^{p+1}$. Then, it can be proven (see Proposition 3.2.3 of Bardis [2]) that $M_2 \circ \mathfrak{h}_{\lambda,1} = e_{\lambda,2}$. Therefore, on U_λ for all λ in Λ,

$$\begin{aligned} L_2 \circ \mathfrak{h}_{\lambda,1} &= e_{\lambda,0} \oplus e_{\lambda,1} \oplus e_{\lambda,2} \oplus \bigoplus_{i=3}^{n} v_\lambda^i \mathfrak{y}_\lambda^{p+1} \\ &= \mathfrak{h}_{\lambda,2}. \end{aligned}$$

Additionally, by Proposition 4.2 of Ru-Stoll [7], L_2 is an isomorphism.

Before going further, more notation needs to be introduced. For each nonnegative integer j, let $s(j) = \frac{j(j-1)}{2} = 0 + 1 + 2 + ... + (j-1)$. Note that for each $j \geq 0$, $s(j) + j = s(j+1)$ and for $j \geq 1$, $s(j-1) + j = s(j) + 1$. Now, choose p in $\mathbb{N}$ such that $p > s(n)$. For $0 \leq j \leq k \leq n$, define γ_{jk} in $\mathfrak{L}_j$ such that

$$\gamma_{jk} = < \mathfrak{g}_0 \wedge ... \wedge \mathfrak{g}_j, \mathfrak{e}_0 \wedge ... \wedge \mathfrak{e}_{j-1} \wedge \mathfrak{e}_k >,$$

or, equivalently, that

$$\gamma_{jk} = \begin{vmatrix} g_{00} & \cdots & g_{0,j-1} & g_{0k} \\ \vdots & & \vdots & \vdots \\ g_{j-1,0} & \cdots & g_{j-1,j-1} & g_{j-1,k} \\ g_{j0} & \cdots & g_{j,j-1} & g_{jk} \end{vmatrix}.$$

Since $\mathfrak{e} = (\mathfrak{e}_0, ..., \mathfrak{e}_n)$ is perfect for $\mathfrak{G}$, then $\gamma_j = \gamma_{jj} \not\equiv 0$ and $\gamma_0 \equiv 1$. Next, set $\rho_0 \equiv 1$ and for each $1 \leq j \leq n$, define $0 \not\equiv \rho_j$ in $\mathfrak{L}_{s(j)}$ such that

$$\rho_j = \gamma_0 \cdot \gamma_1 \cdots \gamma_{j-1}.$$

In addition, set $\rho_{j0} = \rho_j$ and $0 \not\equiv \rho_{ji} = \gamma_i \cdot \gamma_{i+1} \cdots \gamma_{j-1}$ in $\mathfrak{L}_{s(j)-s(i)}$ for $0 \le i < j \le n$. Here, $\rho_{j0} = \rho_j = \rho_i \rho_{ji}$. Next, for $0 \le i \le j \le n$, denote

$$h_{ji} = (-1)^{j+i} < \mathfrak{g}_0 \wedge \ldots \wedge \mathfrak{g}_{i-1} \wedge \mathfrak{g}_{i+1} \wedge \ldots \wedge \mathfrak{g}_j, \mathfrak{e}_0 \wedge \ldots \wedge \mathfrak{e}_{j-1} > .$$

Notice that $h_{jj} = \gamma_{j-1,j-1} = \gamma_{j-1} \not\equiv 0$. Then, for $0 \le i < j \le m \le n$, define k_{ji} in $\mathfrak{L}_{s(j)-s(i)}$ and $\mathfrak{u}_{jm}$ in $\mathfrak{L}_{s(j)+1}$ by

$$\begin{aligned} k_{ji} &= h_{ji}\rho_{j-1,i} \quad \text{and} \\ \mathfrak{u}_{jm} &= \rho_{j-1}\gamma_{jm}. \end{aligned}$$

Also, set $\mathfrak{u}_j = \mathfrak{u}_{jj} = \rho_{j-1}\gamma_{jj} \not\equiv 0$.

Define the following linear maps for $0 \le i \le j \le n$:

$$\begin{aligned} P_j &= \mathbf{F}[s(j)+1, p-s(j)] : Y(p+1) \to Y(p-s(j)); \\ A_j &= \mathbf{F}_{k_{j0}}[s(j), p-s(j)] : Y(p) \to Y(p-s(j)); \\ B_{ji} &= \mathbf{F}_{k_{ji}}[s(j)-s(i), p-s(j)] : Y(p-s(i)) \to Y(p-s(j)); \\ \Phi_j &= \tilde{\mathbf{F}}_{\mathfrak{u}_j}[s(j)+1, p-s(j)] : Y(p+1) \to Y(p+1); \\ \Psi_{ji} &= \mathbf{F}_{\mathfrak{u}_{ji}}[s(j)+1, p-s(j)] : Y(p+1) \to Y(p-s(j)). \end{aligned}$$

Here, Φ_j is an isomorphism. Now, the induction will be completed. Select $\mathfrak{z} = z_0 \oplus \ldots \oplus z_n$ in $Z(p)$. For $1 \le j \le n$, define the linear maps $M_j : Z(p) \to Y(p+1)$ and $L_j : Z(p) \to Z(p)$ by

$$\begin{aligned} M_j(\mathfrak{z}) &= -A_j(z_0) - \sum_{i=1}^{j-1} B_{ji}(P_i(z_i)) + \Phi_j(z_j) + \sum_{i=j+1}^{n} \Psi_{ji}(z_i); \\ L_j(\mathfrak{z}) &= z_0 \oplus z_1 \oplus \ldots \oplus z_{j-1} \oplus M_j(z) \oplus z_{j+1} \oplus \ldots \oplus z_n. \end{aligned}$$

By Lemma 4.4 of Ru-Stoll [7], L_j is an isomorphism for $0 \le j \le n$. Additionally, let $L_\alpha = L_n \circ L_{n-1} \circ \ldots \circ L_1 \circ L_0 : Z(p) \to Z(p)$. Then, L_α is also a linear isomorphism. As denoted previously, $e_{\lambda 0} : U_\lambda \to Y(p)$ is a holomorphic map for λ in Λ and is given by

$$e_{\lambda 0} = < \mathfrak{v}_\lambda, \mathfrak{g}_0 > \Delta_\lambda \mathfrak{y}_\lambda^p. \tag{3.9}$$

Also, for $1 \le j \le n$ and for all λ in Λ, define $\mathfrak{x}_{\lambda j} : U_\lambda \to X(s(j)+1, p-s(j)) \subseteq Y(p+1)$ and $e_{\lambda j} : U_\lambda \to Y(p+1)$ by

$$\begin{aligned} \mathfrak{x}_{\lambda j} &= \mathfrak{x}_{\lambda \mathfrak{u}_j}[s(j)+1, p-s(j)]; \\ e_{\lambda j} &= < \mathfrak{v}_\lambda, \mathfrak{g}_j > \rho_j \Delta_\lambda^{s(j)+1} \mathfrak{y}_\lambda^{p-s(j)} + v_\lambda^j \mathfrak{x}_{\lambda j}. \end{aligned} \tag{3.10}$$

For all λ in Λ, write $\mathfrak{h}_{\lambda,-1} = \mathfrak{h}_\lambda$ on U_λ. Then, for $0 \le j \le n$, define a holomorphic map $\mathfrak{h}_{\lambda j} : U_\lambda \to Z(p)$ by

$$\mathfrak{h}_{\lambda,j} = e_{\lambda,0} \oplus e_{\lambda,1} \oplus \ldots \oplus e_{\lambda,j} \oplus \bigoplus_{i=j+1}^{n} v_\lambda^i \mathfrak{y}_\lambda^{p+1}.$$

In the cases when $j = 1$ and $j = 2$, the definitions for M_j, L_j, and $\mathfrak{h}_{\lambda,j}$ coincide with the previous ones. Then, for $0 \le j \le n$ and on U_λ for all λ in Λ,

$$\begin{aligned} M_j \circ \mathfrak{h}_{\lambda,j-1} &= e_{\lambda j} \quad \text{and} \\ L_j \circ \mathfrak{h}_{\lambda,j-1} &= \mathfrak{h}_{\lambda,j}. \end{aligned}$$

See Proposition 3.2.5 of Bardis [2].

For all λ in Λ, the modified p^{th} Steinmetz map $\mathfrak{h}^{\alpha}_{[p],\lambda} : U_\lambda \to Z(p)$ is defined to be $\mathfrak{h}^{\alpha}_{[p],\lambda} = \mathfrak{h}^{\alpha}_{\lambda} = \mathfrak{h}_{\lambda,n}$ for $\alpha : \mathbb{Z}[0,n] \to \mathfrak{G}$ selected above. For α in $\mathfrak{T}$ and for all λ in Λ,

$$\mathfrak{h}^{\alpha}_{\lambda} = L_\alpha \circ \mathfrak{h}_\lambda.$$

Here, $\mathfrak{h}^{\alpha}_{\lambda} = e_{\lambda 0} \oplus ... \oplus e_{\lambda n} = (e_{\lambda 0}, ..., e_{\lambda n})$. Also, $\mathfrak{h}^{\alpha}_{\lambda}$ is linearly non-degenerated and defines a meromorphic map $h^{\alpha}_{[p]} = h^{\alpha} : M \to \mathbb{P}(Z(p))$ such that
$h^{\alpha}|U_\lambda : U_\lambda \to \mathbb{P}(Z(p))$ is given by

$$h^{\alpha}|U_\lambda = \mathbb{P}(\mathfrak{h}^{\alpha}_{\lambda}).$$

See Corollary 3.2.6 of Bardis [2].

Chapter 4

4.1 Important lemmata

The following important assumptions (C1)-(C14) must be introduced:

ASSUMPTIONS C

(C1)-(C8) are the same as (A1)-(A8).

(C9) $\mathfrak{G}$ is a finite set of meromorphic maps $g : M \to \mathbb{P}(V^*)$ such that at least one g in $\mathfrak{G}$ is nonconstant and $\#\mathfrak{G} \geq n+1$.

(C10) $\mathfrak{G}$ is in general position.

(C11) $\mathfrak{K}_{\mathfrak{G}}$ is the field generated by the coordinates of all g in $\mathfrak{G}$.

(C12) f is linearly non-degenerated over $\mathfrak{K}_{\mathfrak{G}}$.

(C13) For each p in $\mathbb{N}$ with $p > \frac{1}{2}n(n+1)$, let the pth Steinmetz map $h_{[p]}$ for $(\mathfrak{G}, f)$ be general for $\mathbb{B}$.

(C14) For each g in $\mathfrak{G}$, g grows slower than f, i.e.

$$\frac{T_g(r,s)}{T_f(r,s)} \to 0 \quad \text{as} \quad r \to \infty.$$

To begin with, another definition is necessary. f is said to have transcendental growth if and only if

$$\lim_{r\to\infty} \frac{T_f(r,s)}{\log r} \quad = \quad \infty.$$

Since there exists a nonconstant g in $\mathfrak{G}$ by (C9) and since each g in $\mathfrak{G}$ grows slower than f by (C14), then f has transcendental growth.

The notations and constructions of Chapter 3 will remain valid. Additionally, it is required that for each λ in Λ, there is a given holomorphic chart

$$\mathfrak{z}_\lambda \quad = \quad (z_{1\lambda}, ..., z_{m\lambda}) : U_\lambda \to U'_\lambda.$$

Then, a holomorphic frame of the canonical bundle $K(M)$ of M over U_λ is given to be

$$\zeta_\lambda \quad = \quad dz_{1\lambda} \wedge ... \wedge dz_{m\lambda}.$$

For (λ, μ) in $\Lambda[1]$, there exists a unique holomorphic function $\Delta_{\lambda\mu} : U_{\lambda\mu} \to \mathbb{C}_*$ such that

$$\zeta_\lambda = \Delta_{\lambda\mu}\zeta_\mu.$$

Note that $\Delta_{\lambda\lambda} = 1 = \Delta_{\lambda\mu}\Delta_{\mu\lambda}$. For (λ, μ, ρ) in $\Lambda[2]$, $\Delta_{\lambda\mu}\Delta_{\mu\rho}\Delta_{\rho\lambda} = 1$ on $U_{\lambda\mu\rho}$.

To begin with, the characteristic functions of f and $h^\alpha_{[p]} = h^\alpha$ will be related.

LEMMA 4.1 *Suppose* [C1]-[C6] *and* [C9]-[C12] *are valid. Take* α *in* $\mathfrak{T}$. *Then, for every* $s > 0$, *there exists a constant* $c(s)$ *such that*

$$T_{h^\alpha_{[p]}}(r, s) \leq T_f(r, s) + (p+1)T_y(r, s) + c(s) \tag{4.1}$$

for all $r > s$. *Hence, as* $r \to \infty$,

$$T_{h^\alpha_{[p]}}(r, s) \leq T_f(r, s) + o(T_f(r, s)).$$

PROOF: See Bardis [2], pp. 105-108.

Now, fix p in $\mathbb{N}$ such that $p > \frac{1}{2}n(n+1)$. Take $\mathfrak{c}$ in $\underset{k(p)+1}{\wedge} Z(p)^*$ such that $\| \mathfrak{c} \| = 1$ where the dimension of $Z(p)$ is $k(p)+1$. On U_λ for all λ in Λ, define the meromorphic function $W_\lambda : U_\lambda \to \mathbb{C}$ by

$$W_\lambda = < \mathfrak{h}_\lambda \wedge D\mathfrak{h}_\lambda \wedge ... \wedge D^{k(p)}\mathfrak{h}_\lambda, \mathfrak{c} >, \tag{4.2}$$

the Wronski determinant of $\mathfrak{h}_\lambda$ with respect to $\mathfrak{c}$. Consequently, since $\mathbb{B}$ is general for $\mathfrak{h}_\lambda : U_\lambda \to Z(p)$, meaning that

$$\mathfrak{h}_\lambda \wedge D\mathfrak{h}_\lambda \wedge ... \wedge D^{k(p)}\mathfrak{h}_\lambda \not\equiv 0,$$

then $W_\lambda \not\equiv 0$.

By 3.8, on $U_{\lambda\mu}$ for (λ, μ) in $\Lambda[1]$,

$$\begin{aligned} W_\lambda &= < \mathfrak{h}_\lambda \wedge D\mathfrak{h}_\lambda \wedge ... \wedge D^{k(p)}\mathfrak{h}_\lambda, \mathfrak{c} > \\ &= < T^{p+1}_{\lambda\mu} v_{\lambda\mu}\mathfrak{h}_\mu \wedge D(T^{p+1}_{\lambda\mu} v_{\lambda\mu}\mathfrak{h}_\mu) \wedge ... \wedge D^{k(p)}(T^{p+1}_{\lambda\mu} v_{\lambda\mu}\mathfrak{h}_\mu), \mathfrak{c} > \\ &= T^{(p+1)(k(p)+1)}_{\lambda\mu} v^{k(p)+1}_{\lambda\mu} W_\mu. \end{aligned} \tag{4.3}$$

For all α in $\mathfrak{T}$ and λ in Λ, define $W^\alpha_\lambda : U_\lambda \to \mathbb{C}$ by

$$W^\alpha_\lambda = < \mathfrak{h}^\alpha_\lambda \wedge D\mathfrak{h}^\alpha_\lambda \wedge ... \wedge D^{k(p)}\mathfrak{h}^\alpha_\lambda, \mathfrak{c} > .$$

Also,

$$\begin{aligned} |W_\lambda| &= \| \mathfrak{h}_\lambda \wedge D\mathfrak{h}_\lambda \wedge ... \wedge D^{k(p)}\mathfrak{h}_\lambda \|, \\ |W^\alpha_\lambda| &= \| \mathfrak{h}^\alpha_\lambda \wedge D\mathfrak{h}^\alpha_\lambda \wedge ... \wedge D^{k(p)}\mathfrak{h}^\alpha_\lambda \| . \end{aligned}$$

There exists an isomorphism $\hat{L}_\alpha : \underset{k(p)+1}{\wedge} Z(p) \to \underset{k(p)+1}{\wedge} Z(p)$ induced by the linear isomorphism $L_\alpha : Z(p) \to Z(p)$ and given by

$$\hat{L}_\alpha(z_0 \wedge ... \wedge z_{k(p)}) = L_\alpha(z_0) \wedge ... \wedge L_\alpha(z_{k(p)})$$

for z_j in $Z(p)$ with $0 \leq j \leq k(p)$. Since the dimension of $\underset{k(p)+1}{\wedge} Z(p)$ is 1 and since $\hat{L}_\alpha$ is an isomorphism, then there exists a Λ_α in $\mathbb{C}_* = \mathbb{C} - \{0\}$ such that

$$\hat{L}_\alpha(\mathfrak{z}) = \Lambda_\alpha \mathfrak{z}$$

for $\mathfrak{z}$ in $\underset{k(p)+1}{\wedge} Z(p)$. Thus, on U_α for all λ in Λ,

$$\begin{aligned} \mathfrak{h}_\lambda^\alpha \wedge D\mathfrak{h}_\lambda^\alpha \wedge \ldots \wedge D^{k(p)+1}\mathfrak{h}_\lambda^\alpha &= \hat{L}_\alpha(\mathfrak{h}_\lambda \wedge D\mathfrak{h}_\lambda \wedge \ldots \wedge D^{k(p)+1}\mathfrak{h}_\lambda) \\ &= \Lambda_\alpha \mathfrak{h}_\lambda \wedge D\mathfrak{h}_\lambda \wedge \ldots \wedge D^{k(p)+1}\mathfrak{h}_\lambda. \end{aligned}$$

This implies that on U_λ for λ in Λ,

$$W_\lambda^\alpha = \Lambda_\alpha W_\lambda.$$

Hence, $W_\lambda^\alpha \not\equiv 0$.

Next, define holomorphic function $0 \not\equiv K_\lambda : U_\lambda \to \mathbb{C}$ for all λ in Λ such that

$$K_\lambda = \left(v_\lambda^0\right)^{q(p)} \left(v_\lambda^1 \cdots v_\lambda^n\right)^{q(p+1)} \Delta_\lambda^{(p+1)(k(p)+1)}. \tag{4.4}$$

Then, on $U_{\lambda\mu} \neq \emptyset$, the resulting transition formula is

$$K_\lambda | U_{\lambda\mu} = v_{\lambda\mu}^{k(p)+1} T_{\lambda\mu}^{(p+1)(k(p)+1)} K_\mu | U_{\lambda\mu}. \tag{4.5}$$

For (λ, μ) in $\Lambda[1]$,

$$\begin{aligned} \frac{W_\lambda}{K_\lambda} | U_{\lambda\mu} &= \frac{v_{\lambda\mu}^{k(p)+1} T_{\lambda\mu}^{(p+1)(k(p)+1)} W_\mu}{v_{\lambda\mu}^{k(p)+1} T_{\lambda\mu}^{(p+1)(k(p)+1)} K_\mu} \\ &= \frac{W_\mu}{K_\mu} | U_{\lambda\mu} . \end{aligned}$$

by 4.3 and 4.5. Thus, there exists a unique meromorphic function $0 \not\equiv S(f) : M \to \mathbb{C}$ given by

$$S(f)|U_\lambda = \frac{W_\lambda}{K_\lambda}.$$

for each λ in Λ.

Over U_λ for each λ in Λ, let $\mathfrak{v}_\lambda$ be the reduced representation of f, $\overset{\triangle}{\mathfrak{v}_\lambda}$ be the associated holomorphic frame of L_f, and u_λ be the holomorphic frame of T. For (λ, μ) in $\Lambda[1]$, the following transition formulas are valid on $U_{\lambda\mu}$:

$$\mathfrak{v}_\lambda = v_{\lambda\mu}\mathfrak{v}_\mu \quad \text{and} \quad \overset{\triangle}{\mathfrak{v}_\lambda} = v_{\mu\lambda}\overset{\triangle}{\mathfrak{v}_\mu} \;;$$

$$\Delta_\lambda = T_{\lambda\mu}\Delta_\mu \quad \text{and} \quad u_\lambda = T_{\mu\lambda}u_\mu.$$

For each λ in Λ, define the meromorphic section

$$\hat{W}_\lambda = W_\lambda \overset{\triangle}{\mathfrak{v}_\lambda}{}^{k(p)+1} \otimes u_\lambda^{(p+1)(k(p)+1)} \not\equiv 0.$$

of the holomorphic line bundle $L_f^{k(p)+1} \otimes T^{(p+1)(k(p)+1)}$ over U_λ. For (λ, μ) in $\Lambda[1]$,

$$\begin{aligned}\hat{W}_\lambda|U_{\lambda\mu} &= W_\lambda \overset{\triangle}{\mathfrak{v}_\lambda}{}^{k(p)+1} \otimes u_\lambda^{(p+1)(k(p)+1)} \\ &= T_{\lambda\mu}^{(p+1)(k(p)+1)} v_{\lambda\mu}^{k(p)+1} W_\mu v_{\mu\lambda}^{k(p)+1} \overset{\triangle}{\mathfrak{v}_\mu}{}^{k(p)+1} \otimes T_{\mu\lambda}^{(p+1)(k(p)+1)} u_\mu^{(p+1)(k(p)+1)} \\ &= W_\mu \overset{\triangle}{\mathfrak{v}_\mu}{}^{k(p)+1} \otimes u_\mu^{(p+1)(k(p)+1)} \\ &= \hat{W}_\mu|U_{\lambda\mu}.\end{aligned}$$

Thus, there exists a unique meromorphic section W of $L_f^{k(p)+1} \otimes T^{(p+1)(k(p)+1)}$ over M such that for all λ in Λ,

$$W|U_\lambda = \hat{W}_\lambda.$$

It is obvious that W is not the null section. Also, the divisor of W denoted μ_W is such that for each λ in Λ,

$$\mu_W|U_\lambda = \mu_{\hat{W}_\lambda} = \mu_{W_\lambda}.$$

Denote N_{μ_W} as N_W.

Define another meromorphic section $\mathbb{K}$ of $L_f^{k(p)+1} \otimes T^{(p+1)(k(p)+1)}$ over M by

$$\mathbb{K} = \frac{1}{S(f)} W \not\equiv 0.$$

For λ in Λ,

$$\begin{aligned}\mathbb{K}|U_\lambda &= \frac{K_\lambda}{W_\lambda} W_\lambda \overset{\triangle}{\mathfrak{v}_\lambda}{}^{k(p)+1} \otimes u_\lambda^{(p+1)(k(p)+1)} \\ &= K_\lambda \overset{\triangle}{\mathfrak{v}_\lambda}{}^{k(p)+1} \otimes u_\lambda^{(p+1)(k(p)+1)}. \end{aligned} \tag{4.6}$$

Because K_λ is holomorphic, $\mathbb{K}$ is a holomorphic section.

Now, $-N_W(r,s)$ will be estimated in terms of $N_\beta(r,s)$.

LEMMA 4.2 *Suppose* [C1]-[C13] *are valid. For* $0 < s < r$,

$$-N_W(r,s) \leq \frac{k(p)[k(p)+1]}{2} N_\beta(r,s). \tag{4.7}$$

PROOF: Since Θ is a holomorphic section of the canonical bundle $K(M)$, then $X = W \otimes \Theta^{\frac{k(p)[k(p)+1]}{2}}$ is a meromorphic section of $L_f^{k(p)+1} \otimes T^{(p+1)(k(p)+1)} \otimes K(M)^{\frac{k(p)[k(p)+1]}{2}}$.

For λ in Λ, $\Theta|U_\lambda = \Theta_\lambda \zeta_\lambda$ and

$$W_\lambda \Theta_\lambda^{\frac{k(p)[k(p)+1]}{2}} = < \mathfrak{h}_\lambda \wedge D\mathfrak{h}_\lambda \wedge \ldots \wedge D^{k(p)}\mathfrak{h}_\lambda \Theta_\lambda^{\frac{k(p)[k(p)+1]}{2}}, \mathfrak{c} >$$

is holomorphic. Thus,

$$X|U_\lambda = W_\lambda \Theta_\lambda^{\frac{k(p)[k(p)+1]}{2}} \overset{\triangle}{\mathfrak{v}_\lambda}{}^{k(p)+1} \otimes u_\lambda^{(p+1)(k(p)+1)} \zeta_\lambda^{\frac{k(p)[k(p)+1]}{2}}$$

is holomorphic. Therefore, $X \not\equiv 0$ must be holomorphic as well. By definition $\beta = \mu_\Theta$ and therefore

$$0 \leq \mu_X = \mu_W + \frac{k(p)[k(p)+1]}{2}\mu_\Theta \qquad \text{on } M$$

or equivalently

$$-\mu_W \leq \frac{k(p)[k(p)+1]}{2}\beta \qquad \text{on } M$$

which implies

$$-N_W(r,s) \leq \frac{k(p)[k(p)+1]}{2}N_\beta.$$

q.e.d.

Recall that $s(j) = \frac{j(j-1)}{2}$ for j in $\mathbb{Z}_+$ and the dimension of $Y(p)$ is $q(p)$. For $0 \leq j \leq n$, denote

$$q_j = q(p - s(j)) \text{ and } r_j = q(p+1) - q_j.$$

Recall as well that $\mathfrak{T}$ is the set all injective mappings from $\mathbb{Z}[0,n]$ to $\mathfrak{G}$. For α in $\mathfrak{T}$ and $0 \leq j \leq n$, set $g_j = \alpha(j)$ as done in the previous chapter. Define a meromorphic function $0 \not\equiv K_\lambda^\alpha : U_\lambda \to \mathbb{C}$ by

$$\begin{aligned} K_\lambda^\alpha &= \Delta_\lambda^{(p+1)[k(p)+1]} \prod_{j=0}^{n} < \mathfrak{v}_\lambda, \mathfrak{g}_j >^{q(p-s(j))} \prod_{j=1}^{n} (v_\lambda^j)^{q(p+1)-q(p-s(j))} \\ &= \Delta_\lambda^{(p+1)[k(p)+1]} \prod_{j=0}^{n} < \mathfrak{v}_\lambda, \mathfrak{g}_j >^{q_j} \prod_{j=1}^{n} (v_\lambda^j)^{r_j}. \end{aligned} \tag{4.8}$$

Take (λ, μ) in $\Lambda[1]$. Then, on $U_{\lambda\mu}$, consider the transition formula for K_λ^α.

$$\begin{aligned} K_\lambda^\alpha | U_{\lambda\mu} &= (T_{\lambda\mu}\Delta_\mu)^{(p+1)[k(p)+1]} \prod_{j=0}^{n} (< v_{\lambda\mu}\mathfrak{v}_\mu, \mathfrak{g}_j >)^{q(p-s(j))} \cdot \\ &\quad \prod_{j=1}^{n} (v_{\lambda\mu} v_\mu^j)^{q(p+1)-q(p-s(j))} \\ &= T_{\lambda\mu}^{(p+1)[k(p)+1]} v_{\lambda\mu}^{q(p)+nq(p+1)} \Delta_\mu^{(p+1)[k(p)+1]} \cdot \\ &\quad \prod_{j=0}^{n} < \mathfrak{v}_\mu, \mathfrak{g}_j >^{q(p-s(j))} \prod_{j=1}^{n} (v_\mu^j)^{q(p+1)-q(p-s(j))} \\ &= T_{\lambda\mu}^{(p+1)[k(p)+1]} v_{\lambda\mu}^{k(p)+1} K_\mu^\alpha | U_{\lambda\mu}. \end{aligned}$$

Additionally, using 4.3 on $U_{\lambda\mu}$,

$$\begin{aligned} W_\lambda^\alpha &= \Lambda_\alpha W_\lambda \\ &= \Lambda_\alpha T_{\lambda\mu}^{(p+1)(k(p)+1)} v_{\lambda\mu}^{k(p)+1} W_\mu \\ &= T_{\lambda\mu}^{(p+1)(k(p)+1)} v_{\lambda\mu}^{k(p)+1} W_\mu^\alpha. \end{aligned}$$

Therefore,

$$\frac{W^\alpha_\lambda}{K^\alpha_\lambda}|U_{\lambda\mu} = \frac{W^\alpha_\mu}{K^\alpha_\mu}|U_{\lambda\mu}$$

Consequently, there exists a u nique meromorphic function $0 \not\equiv S_\alpha(f) : M \to \mathbb{C}$ such that

$$S_\alpha(f)|U_\lambda = \frac{W^\alpha_\lambda}{K^\alpha_\lambda} \tag{4.9}$$

for all λ in Λ.

Now, let $\mathfrak{b}_1, \mathfrak{b}_2, ..., \mathfrak{b}_{q(p)}$ be an orthonormal base of $Y(p)$. Since $\mathfrak{y}^p_\lambda : U_\lambda \to Y(p)$, then there exists unique holomorphic functions $B^1_\lambda, ..., B^{q(p)}_\lambda$ on U_λ for all λ in Λ such that

$$\mathfrak{y}^p_\lambda = \sum_{i=1}^{q(p)} B^i_\lambda \mathfrak{b}_i. \tag{4.10}$$

By 3.5, $B^i_\lambda = T^p_{\lambda\mu} B^i_\mu$ for (λ, μ) in $\Lambda[1]$. Therefore, on $U_{\lambda\mu}$,

$$\frac{B^i_\lambda}{\Delta^p_\lambda} = \frac{T^p_{\lambda\mu} B^i_\mu}{T^p_{\lambda\mu}\Delta^p_\mu} = \frac{B^i_\mu}{\Delta^p_\mu}.$$

Define global meromorphic functions $\phi_i : M \to \mathbb{C}$ for $1 \le i \le q(p)$ by

$$\phi_i|U_\lambda = \frac{B^i_\lambda}{\Delta^p_\lambda} \tag{4.11}$$

for all λ in Λ. By Theorem 3.9 of Stoll [14], $\phi_1, ..., \phi_{q(p)}$ form a base for $\mathfrak{L}_p$.

Now, consider the direct sum decomposition of $Y(p+s)$ given on p. 122 of Ru-Stoll [7]. For $1 \le j \le n$,

$$Y(p+1) = Y(p - s(j)) \oplus X(s(j)+1, p-s(j)), \tag{4.12}$$

where the dimension of $Y(p+1)$ is $q(p+1) = r_j + q_j$. Then, an orthonormal base of $Y(p+1)$ can be written as $\mathfrak{b}_{j1}, ..., \mathfrak{b}_{jq_j}, \mathfrak{i}_{j1}, ..., \mathfrak{i}_{jr_j}$ in which the first q_j vectors form a base for $Y(p - s(j))$ and the next r_j vectors form a base for $X(s(j)+1, p-s(j))$. In turn, there exists unique holomorphic functions $B^{j1}_\lambda, ..., B^{jq_j}_\lambda$ and $X^{j1}_\lambda, ..., X^{jr_j}_\lambda$ on U_λ for all λ in Λ such that

$$\mathfrak{y}^{p+1}_\lambda = \sum_{i=1}^{q_j} B^{ji}_\lambda \mathfrak{b}_{ji} + \sum_{k=1}^{r_j} X^{jk}_\lambda \mathfrak{i}_{jk}. \tag{4.13}$$

Again, by Theorem 3.9 of Stoll [14], global meromorphic functions $\chi_{j1}, ..., \chi_{jq_j}, \psi_{j1}, ..., \psi_{jr_j}$ yielding a base for $\mathfrak{L}_{p+1}$ can be defined by

$$\chi_{ji}|U_\lambda = \frac{B^{ji}_\lambda}{\Delta^{p+1}_\lambda}$$

for $j = 1, ..., q_j$ and

$$\psi_{jk}|U_\lambda = \frac{X_\lambda^{jk}}{\Delta_\lambda^{p+1}}$$

for $j = 1, ..., r_j$. On U_λ for all λ in Λ, the direct sum decomposition 4.12 of $Y(p+1)$ implies that

$$\mathfrak{y}_\lambda^{p+1} = \Delta_\lambda^{s(j)+1}\mathfrak{y}_\lambda^{p-s(j)} \oplus \mathfrak{x}_\lambda[s(j)+1, p-s(j)] \tag{4.14}$$

and thus

$$\Delta_\lambda^{s(j)+1}\mathfrak{y}_\lambda^{p-s(j)} = \sum_{i=1}^{q_j} B_\lambda^{ji}\mathfrak{b}_{ji} \quad \text{or}$$

$$\mathfrak{y}_\lambda^{p-s(j)} = \sum_{i=1}^{q_j} \frac{B_\lambda^{ji}}{\Delta_\lambda^{s(j)+1}}\mathfrak{b}_{ji}. \tag{4.15}$$

This implies that $\tilde{B}_\lambda^{ji} = \frac{B_\lambda^{ji}}{\Delta_\lambda^{s(j)+1}}$ must be holomorphic on U_λ for λ in Λ and $1 \leq j \leq q_j$. Hence, for $1 \leq j \leq q_j$,

$$\chi_{ji}|U_\lambda = \frac{\tilde{B}_\lambda^{ji}}{\Delta_\lambda^{p-s(j)}} = \frac{B_\lambda^{ji}}{\Delta_\lambda^{p+1}} \tag{4.16}$$

form a base for $\mathfrak{L}_{p-s(j)}$ as a linear subspace of $\mathfrak{L}_{p+1}$.

Also, for $1 \leq j \leq q_j$ and λ in Λ, since

$$\mathfrak{x}_{\lambda j} = \mathfrak{x}_\lambda[s(j)+1, p-s(j)] : U_\lambda \to X(s(j)+1, p-s(j)) \subseteq Y(p+1),$$

4.12, 4.13, and 4.14 imply

$$\mathfrak{x}_{\lambda j} = \sum_{k=1}^{r_j} X_\lambda^{jk}\mathfrak{i}_{jk} \tag{4.17}$$

for $X_\lambda^{jk} : U_\lambda \to \mathbb{C}$ holomorphic functions.

Now, suppose that $\mathfrak{b}_{j1}^*, ..., \mathfrak{b}_{jq_j}^*, \mathfrak{i}_{j1}^*, ..., \mathfrak{i}_{jr_j}^*$ form the base of $\mathfrak{L}_{p+1}$ with its dual base $\mathfrak{b}_{j1}, ..., \mathfrak{b}_{jq_j}, \mathfrak{i}_{j1}, ..., \mathfrak{i}_{jr_j}$ of $Y(p+1)$. Hence, $\psi_{jk} \in \mathfrak{L}_{p+1}$ are given by Lemma 3.1.15 in Bardis [2] such that

$$X_\lambda^{jk} = <\mathfrak{x}_{\lambda j}, \mathfrak{i}_{jk}^*> = \Delta_\lambda^{p+1}\psi_{jk}. \tag{4.18}$$

An orthonormal base of $Z(p)$ is

$$\mathfrak{b} = \{\mathfrak{b}_1, ..., \mathfrak{b}_{\mathfrak{q}_0}, \mathfrak{b}_{11}, ..., \mathfrak{b}_{1\mathfrak{q}_1}, \mathfrak{i}_{11}, ..., \mathfrak{i}_{1\mathfrak{r}_1}, ..., \mathfrak{b}_{\mathfrak{n}1}, ..., \mathfrak{b}_{\mathfrak{n}\mathfrak{q}_\mathfrak{n}}, \mathfrak{i}_{\mathfrak{n}1}, ..., \mathfrak{i}_{\mathfrak{n}\mathfrak{r}_\mathfrak{n}}\}$$

since $Z(p) = Y(p) \oplus Y(p+1)^n$. By multiplying $\mathfrak{b}_1$ by a complex number in the unit circle, it can be assumed without loss of generality that $\mathfrak{c}$ equals the exterior product of these base vectors.

Next, the Lemma of the Logarithmic Derivative will be applied to $\mathfrak{h}_\lambda^\alpha = (e_{\lambda 0}, ..., e_{\lambda n})$. By 3.9 and 4.10, $e_{\lambda 0} : U_\lambda \to Y(p)$ can be expressed as

$$\begin{aligned} e_{\lambda 0} &= < \mathfrak{v}_\lambda, \mathfrak{g}_0 > \Delta_\lambda \mathfrak{y}_\lambda^p \\ &= \sum_{i=1}^{q_0} < \mathfrak{v}_\lambda, \mathfrak{g}_0 > \Delta_\lambda B_\lambda^i \mathfrak{b}_i. \end{aligned}$$

In addition, by 3.10, 4.15, and 4.17, $e_{\lambda j} : U_\lambda \to Y(p+1)$ can be expressed by

$$\begin{aligned} e_{\lambda j} &= < \mathfrak{v}_\lambda, \mathfrak{g}_j > \rho_j \Delta_\lambda^{s(j)+1} \mathfrak{y}_\lambda^{p-s(j)} + v_\lambda^j \mathfrak{x}_{\lambda j} \\ &= \sum_{i=1}^{q_j} < \mathfrak{v}_\lambda, \mathfrak{g}_j > \rho_j \Delta_\lambda^{s(j)+1} \tilde{B}_\lambda^{ji} \mathfrak{b}_{ji} + \sum_{k=1}^{r_j} v_\lambda^j X_\lambda^{jk} \mathfrak{i}_{jk} \end{aligned}$$

for $1 \leq j \leq n$. To apply the Lemma of the Logarithmic Derivative, the product of the holomorphic coordinate functions P_λ^α needs to be computed with respect to the base given above. Let $P_\lambda^{\alpha j}$ be the product of the coordinates of each $e_{\lambda j}$ on U_λ for all λ in Λ and thus $P_\lambda^\alpha = P_\lambda^{\alpha 0} P_\lambda^{\alpha 1} \cdots P_\lambda^{\alpha n}$.

The definition of $e_{\lambda 0}$ and 4.11 imply that

$$\begin{aligned} P_\lambda^{\alpha 0} &= (< \mathfrak{v}_\lambda, \mathfrak{g}_0 > \Delta_\lambda)^{q_0} B_\lambda^1 \cdots B_\lambda^{q_0} \\ &= < \mathfrak{v}_\lambda, \mathfrak{g}_0 >^{q_0} \Delta_\lambda^{(q_0 + q_0 p)} \prod_{i=1}^{q_0} \phi_i \end{aligned}$$

on U_λ for all λ in Λ. In turn, the definition of $e_{\lambda j}$, 4.16, and 4.18 for $1 \leq j \leq n$ give us that

$$\begin{aligned} P_\lambda^{\alpha j} &= (< \mathfrak{v}_\lambda, \mathfrak{g}_j > \rho_j \Delta_\lambda^{s(j)+1})^{q_j} (v_\lambda^j)^{r_j} \prod_{i=1}^{q_j} \tilde{B}_\lambda^{ji} \prod_{k=1}^{r_j} X_\lambda^{jk} \\ &= < \mathfrak{v}_\lambda, \mathfrak{g}_j >^{q_j} (v_\lambda^j)^{r_j} \Delta_\lambda^{(p+1)q(p+1)} \rho_j^{q_j} \prod_{i=1}^{q_j} \chi_{ji} \prod_{k=1}^{r_j} \psi_{jk} \end{aligned}$$

on U_λ for all λ in Λ.

Thus, for (λ, μ) in $\Lambda[1]$ and for $j > 0$,

$$\begin{aligned} P_\lambda^{\alpha 0} | U_{\lambda\mu} &= v_{\lambda\mu}^{q_0} T_{\lambda\mu}^{(p+1)q_0} P_\mu^{\alpha 0} | U_{\lambda\mu} \\ &= v_{\lambda\mu}^{q(p)} T_{\lambda\mu}^{(p+1)q(p)} P_\mu^{\alpha 0} | U_{\lambda\mu}; \\ P_\lambda^{\alpha j} | U_{\lambda\mu} &= v_{\lambda\mu}^{q_j} v_{\lambda\mu}^{r_j} T_{\lambda\mu}^{(p+1)q(p+1)} P_\mu^{\alpha j} | U_{\lambda\mu} \\ &= v_{\lambda\mu}^{q(p+1)} T_{\lambda\mu}^{(p+1)q(p+1)} P_\mu^{\alpha j} | U_{\lambda\mu}; \\ P_\lambda^{\alpha} | U_{\lambda\mu} &= v_{\lambda\mu}^{q(p)+nq(p+1)} T_{\lambda\mu}^{(p+1)[q(p)+nq(p+1)]} P_\mu^{\alpha} | U_{\lambda\mu} \\ &= v_{\lambda\mu}^{k(p)+1} T_{\lambda\mu}^{(p+1)[k(p)+1]} P_\mu^{\alpha} | U_{\lambda\mu}. \end{aligned}$$

Consequently, on $U_{\lambda\mu}$ for (λ, μ) in $\Lambda[1]$,

$$\begin{aligned} \frac{W_\lambda^\alpha}{P_\lambda^\alpha} &= \frac{v_{\lambda\mu}^{k(p)+1} T_{\lambda\mu}^{(p+1)[k(p)+1]} W_\mu^\alpha}{v_{\lambda\mu}^{k(p)+1} T_{\lambda\mu}^{(p+1)[k(p)+1]} P_\mu^\alpha} \\ &= \frac{W_\mu^\alpha}{P_\mu^\alpha}. \end{aligned}$$

Thus, there exists a unique meromorphic function $\hat{U}_\alpha \not\equiv 0$ on M such that for λ in Λ,

$$\hat{U}_\alpha | U_\alpha = \frac{W^\alpha_\lambda}{P^\alpha_\lambda}.$$

LEMMA 4.3 *Suppose* [C1]-[C6] *and* [C9]-[C13] *are valid. Take α in $\mathfrak{T}$ and $0 < s < r$. Then,*

$$\int_{M<r>} \log^+ |\hat{U}_\alpha|\mathbf{Z}^{\frac{k(p)[k(p)+1]}{2}} \sigma \;\underset{\cdot\cdot}{\leq}\; 7k(p)^2(k(p)+1)\varsigma[\log^+ T_f(r,s) + \log^+ Y_{\mathbb{B}}(r) + \log^+ Ric_\tau(r,s) + \log^+ E(r) + \log^+ r].$$

PROOF: By 4.1, there exist constants $c_j(s)$ for $j = 1, 2$ such that

$$\begin{aligned} T_{h^\alpha_{[p]}}(r,s) &\leq T_f(r,s) + (p+1)T_y(r,s) + c_1(s); \\ \log^+ T_{h^\alpha_{[p]}}(r,s) &\leq \log^+ T_f(r,s) + \log^+ T_y(r,s) + c_2(s) \\ &\underset{\cdot\cdot}{\leq} (1 + \frac{1}{3}) \log^+ T_f(r,s). \end{aligned}$$

Consequently,

$$\begin{aligned} (1 + \frac{1}{2}) \log^+ T_{h^\alpha_{[p]}}(r,s) &\underset{\cdot\cdot}{\leq} (1 + \frac{1}{2})(1 + \frac{1}{3}) \log^+ T_f(r,s) \\ &= 2 \log^+ T_f(r,s). \end{aligned}$$

Now, Theorem 2.12 will be applied via the following translation table:

THM 2.12	V	n	$\mathfrak{b}$	f	$\mathfrak{v}_\lambda$	ϵ	$W_{f,\beta}$
LEM 4.3	$Z(p)$	$k(p)$	$\mathfrak{b}$	h^α	$\mathfrak{h}^\alpha_\lambda$	$\frac{1}{2}$	$\hat{U}_\alpha$

Therefore,

$$\begin{aligned} \int_{M<r>} \log^+ |\hat{U}_\alpha|\mathbf{Z}^{\frac{n(n+1)}{2}} \sigma \;\underset{\cdot\cdot}{\leq}\; & \frac{1}{2}k(p)^2(k(p)+1)(1+\frac{1}{2})\varsigma[7\log^+ T_{h^\alpha}(r,s) + \\ & \log^+ Y_{\mathbb{B}}(r) + 7\log^+ r] + \\ & \frac{7}{2}k(p)^2(k(p)-1)(1+\frac{1}{2})\varsigma[\log^+ Ric_\tau(r,s) + \\ & \log^+ E(r)] \\ \underset{\cdot\cdot}{\leq}\; & 7k(p)^2(k(p)+1)\varsigma[\log^+ T_f(r,s) + \log^+ Y_{\mathbb{B}}(r) + \\ & \log^+ Ric_\tau(r,s) + \log^+ E(r) + \log^+ r]. \end{aligned}$$

q.e.d.

Now, define the meromorphic function $G_\alpha : M \to \mathbb{C}$ by

$$G_\alpha = \prod_{i=1}^{q_0} \phi_i \prod_{j=1}^{n} \left(\rho_j^{q_j} \prod_{i=1}^{q_j} \chi_{ji} \prod_{k=1}^{r_j} \psi_{jk} \right). \tag{4.19}$$

Since $\mathfrak{L}_p$ is contained in $\mathfrak{K}_{\mathfrak{G}}$ for all p in $\mathbb{N}$, then G_α is contained in $\mathfrak{K}_{\mathfrak{G}}$.

Now, for r in $\mathfrak{E}_\tau$, define

$$\hat{m}_{S_\alpha(f)}(r) = \int_{M<r>} \log\sqrt{1+\mathbf{Z}^{k(p)[k(p)+1]}|S_\alpha(f)|^2}\sigma.$$

Denote

$$c_0(p) = 7k(p)^2(k(p)+1)\varsigma.$$

Also, notice that $\log^+ r = o(T_f(r,s))$ since f is transcendental.

LEMMA 4.4 [LEMMA OF THE LOGARITHMIC DERIVATIVE] *Suppose* [C1]-[C6] *and* [C9]-[C13] *are valid. For* α *in* $\mathfrak{T}$ *and* $0 < s < r$, *the following estimate holds:*

$$\hat{m}_{S_\alpha(f)}(r) \mathrel{\underset{\cdot\cdot}{\leq}} c_0(p)[\log^+ Ric_\tau(r,s) + \log^+ E(r) + \log^+ Y_{\mathbb{B}}(r)] + o(T_f(r,s)). \quad (4.20)$$

PROOF: Since $(q_0+q_0p)+n(p+1)q(p+1) = (p+1)[q_0+nq(p+1)] = (p+1)[k(p)+1]$ and $r_j = q(p+1)-q(j) = q(p+1)-q(p-s(j))$, 4.8 and 4.19 yield that $P_\lambda^\alpha = G_\alpha K_\lambda^\alpha$ which implies that

$$S_\alpha(f) = \frac{W_\lambda^\alpha}{K_\lambda^\alpha} = G_\alpha\left(\frac{W_\lambda^\alpha}{P_\lambda^\alpha}\right) = G_\alpha \hat{U}_\alpha$$

on U_λ for all λ in Λ.

Then,

$$\begin{aligned}
\hat{m}_{S_\alpha(f)}(r) &= \int_{M<r>} \log\sqrt{1+\mathbf{Z}^{k(p)[k(p)+1]}|S_\alpha(f)|^2}\sigma \\
&\mathrel{\underset{\cdot\cdot}{\leq}} \int_{M<r>} \log^+ \mathbf{Z}^{\frac{k(p)[k(p)+1]}{2}}|S_\alpha(f)|\sigma + \varsigma\frac{\log 2}{2} \\
&= \int_{M<r>} \log^+ \mathbf{Z}^{\frac{k(p)[k(p)+1]}{2}}|G_\alpha\hat{U}_\alpha| + \varsigma\frac{\log 2}{2} \\
&\mathrel{\underset{\cdot\cdot}{\leq}} \int_{M<r>} \log^+ \mathbf{Z}^{\frac{k(p)[k(p)+1]}{2}}|\hat{U}_\alpha|\sigma + \int_{M<r>} \log^+|G_\alpha|\sigma + \varsigma\frac{\log 2}{2} \\
&\mathrel{\underset{\cdot\cdot}{\leq}} c_0(p)[\log^+ Ric_\tau(r,s) + \log^+ E(r) + \log^+ Y_{\mathbb{B}}(r)] + \\
&\quad\; m_{G_\alpha}(r) + o(T_f(r,s)) \\
&= c_0(p)[\log^+ Ric_\tau(r,s) + \log^+ E(r) + \log^+ Y_{\mathbb{B}}(r)] + \\
&\quad\; o(T_f(r,s)),
\end{aligned}$$

where the last statement follows since G_α is contained in $\mathfrak{K}_{\mathfrak{G}}$ by definition. In particular, by the First Main Theorem and Proposition 6.2 of Stoll [14],

$$\begin{aligned}
m_{G_\alpha}(r) &\leq T_{G_\alpha}(r,s) + o(T_f(r,s)) \\
&\leq o(T_f(r,s)).
\end{aligned}$$

Hence,

$$\begin{aligned}
\hat{m}_{S_\alpha(f)}(r) &\mathrel{\underset{\cdot\cdot}{\leq}} c_0(p)[\log^+ Ric_\tau(r,s) + \log^+ E(r) + \log^+ Y_{\mathbb{B}}(r)] + \\
&\quad\; o(T_f(r,s)).
\end{aligned}$$

q.e.d.

4.2 Main identity

Recall that on U_λ for all λ in Λ,

$$W_\lambda^\alpha = \Lambda_\alpha W_\alpha.$$

THEOREM 4.5 [MAIN IDENTITY] *Suppose* [C1]-[C6] *and* [C9]-[C13] *are valid. Take* α *in* $\mathfrak{T}$ *and let* $g_j = \alpha(j)$ *in* $\mathfrak{G}$ *for* $0 \le j \le n$. *Suppose that* $\mathfrak{e} = (\mathfrak{e}_0, ..., \mathfrak{e}_n)$ *is an orthonormal base of* V *perfect for* $\mathfrak{G}$. *Let* $\mathfrak{a} = (\mathfrak{a}_0, ..., \mathfrak{a}_n)$ *be its dual base. Write* $e_j = \mathbb{P}(\mathfrak{e}_j)$ *and* $a_j = \mathbb{P}(\mathfrak{a}_j)$ *for* $0 \le j \le n$. *Then, the following identity holds*

$$\left|\frac{S_\alpha(f)}{S(f)}\right| = |\Lambda_\alpha| \prod_{j=0}^{n} \left(\frac{\square f, a_j \square\square e_0, g_j \square}{\square f, g_j \square}\right)^{q(p-s(j))}. \tag{4.21}$$

PROOF: Recall that $w_\lambda^{ji} =< \mathfrak{e}_i, \mathfrak{w}_\lambda^j >$ and on U_λ,

$$\mathfrak{g}_j = \frac{\mathfrak{w}_\lambda^j}{< \mathfrak{e}_0, \mathfrak{w}_\lambda^j >}.$$

Also, by 4.4 and 4.8,

$$K_\lambda =< \mathfrak{v}_\lambda, \mathfrak{a}_0 >^{q(p)} \prod_{j=1}^{n} < \mathfrak{v}_\lambda, \mathfrak{a}_j >^{q(p+1)} \Delta_\lambda^{(p+1)(k(p)+1};$$

$$K_\lambda^\alpha =< \mathfrak{v}_\lambda, \mathfrak{g}_0 >^{q(p)} \prod_{j=1}^{n} < \mathfrak{v}_\lambda, \mathfrak{g}_j >^{q(p-s(j))} \prod_{j=1}^{n} < \mathfrak{v}_\lambda, \mathfrak{a}_j >^{q(p+1)-q(p-s(j))} \Delta_\lambda^{(p+1)(k(p)+1}.$$

Recall that $s(0) = 0$. Therefore, for λ in Λ,

$$\begin{aligned}
\frac{S_\alpha(f)}{S(f)} |U_\lambda &= \frac{W_\lambda^\alpha K_\lambda}{W_\lambda K_\lambda^\alpha} \\
&= \frac{\Lambda_\alpha W_\lambda K_\lambda}{W_\lambda K_\lambda^\alpha} \\
&= \Lambda_\alpha \prod_{j=0}^{n} \left(\frac{< \mathfrak{v}_\lambda, \mathfrak{a}_j >}{< \mathfrak{v}_\lambda, \mathfrak{g}_j >}\right)^{q(p-s(j))} \\
&= \Lambda_\alpha \prod_{j=0}^{n} \left(\frac{< \mathfrak{v}_\lambda, \mathfrak{a}_j >< \mathfrak{e}_0, \mathfrak{w}_\lambda^j >}{< \mathfrak{v}_\lambda, \mathfrak{w}_\lambda^j >}\right)^{q(p-s(j))}.
\end{aligned}$$

Consequently,

$$\begin{aligned}
\left|\frac{S_\alpha(f)}{S(f)}\right| |U_\lambda &= |\Lambda_\alpha| \prod_{j=0}^{n} \left(\frac{| < \mathfrak{v}_\lambda, \mathfrak{a}_j > |}{\| \mathfrak{v}_\lambda \| \| \mathfrak{a}_j \|} \frac{| < \mathfrak{e}_0, \mathfrak{w}_\lambda^j > |}{\| \mathfrak{e}_0 \| \| \mathfrak{w}_\lambda^j \|} \frac{\| \mathfrak{v}_\lambda \| \| \mathfrak{w}_\lambda^j \|}{| < \mathfrak{v}_\lambda, \mathfrak{w}_\lambda^j > |}\right)^{q(p-s(j))} \\
&= |\Lambda_\alpha| \prod_{j=0}^{n} \left(\frac{\square f, a_j \square\square e_0, g_j \square}{\square f, g_j \square}\right)^{q(p-s(j))} |U_\lambda.
\end{aligned}$$

Hence,

$$\left|\frac{S_\alpha(f)}{S(f)}\right| = |\Lambda_\alpha| \prod_{j=0}^{n} \left(\frac{\square f, a_j\square\square e_0, g_j\square}{\square f, g_j\square}\right)^{q(p-s(j))}.$$

q.e.d.

Define the global function $R_\alpha : M \to \mathbb{R}_+$ by

$$R_\alpha = \frac{1}{|\Lambda_\alpha|}\mathbf{Z}^{\frac{k(p)[k(p)+1]}{2}} |S_\alpha(f)| \prod_{j=0}^{n} \left(\frac{\square f, a_j\square^{r_j - r_0}}{\square e_0, g_j\square^{q_j}\square f, g_j\square^{r_j - r_0}}\right). \tag{4.22}$$

Notice here that $r_j - r_0 = q_0 - q_j = q(p) - q(p - s(j)) \geq 0$.

Also, define the function $\Xi : M \to \mathbb{R}_+$ by

$$\Xi = \frac{1}{\mathbf{Z}^{\frac{k(p)[k(p)+1]}{2}} |S(f)|} \prod_{j=0}^{n} \left(\frac{1}{\square f, a_j\square}\right)^{q(p)}. \tag{4.23}$$

Then, by using 4.21, 4.22, and 4.23, the Main Identity can be reformulated in the following manner:

$$\begin{aligned} R_\alpha \Xi &= \prod_{j=0}^{n} \left(\frac{\square f, a_j\square\square e_0, g_j\square}{\square f, g_j\square}\right)^{q_j} \prod_{j=0}^{n} \frac{\square f, a_j\square^{q_0 - q_j}}{\square e_0, g_j\square^{q_j}\square f, g_j\square^{q_0 - q_j}} \prod_{j=0}^{n} \frac{1}{\square f, a_j\square^{q_0}} \\ &= \prod_{j=0}^{n} \left(\frac{1}{\square f, g_j\square}\right)^{q(p)}. \end{aligned}$$

Also, define the function $Q_\alpha : M \to \mathbb{R}_+$ by

$$\begin{aligned} Q_\alpha &= \prod_{j=0}^{n} \frac{1}{\square f, g_j\square} \\ &= [R_\alpha \Xi]^{\frac{1}{q(p)}}. \end{aligned}$$

4.3 Functions θ_α and Λ

Define the function $\theta_\alpha : \mathbb{R}^+ \to \mathbb{R}_+$ by

$$\theta_\alpha(r) = \int_{M\langle r\rangle} \log^+ R_\alpha \sigma. \tag{4.24}$$

LEMMA 4.6 *Suppose* [C1]-[C6] *and* [C9]-[C14] *are valid. Take* α *in* $\mathfrak{T}$ *and* $0 < s < r$. *Then, as* $r \to \infty$,

$$\begin{aligned} \theta_\alpha(r) \leq\ & \sum_{j=1}^{n} (r_j - r_0) T_f(r, s) + c_0(p)[\log^+ Ric_\tau(r, s) + \log^+ E(r) + \\ & \log^+ Y_{\mathbb{B}}(r)] + o(T_f(r, s)). \end{aligned} \tag{4.25}$$

PROOF: From 4.22, it can be concluded that

$$R_\alpha \leq \frac{1}{|\Lambda_\alpha|}\sqrt{1+\mathbf{Z}^{k(p)[k(p)+1]}|S_\alpha(f)|^2}\prod_{j=0}^{n}\frac{1}{\square f, g_j \square^{r_j-r_0}}\prod_{j=0}^{n}\frac{1}{\square e_0, g_j \square^{q_j}}$$

which implies that

$$\begin{aligned}\theta_\alpha(r) &\leq \hat{m}_{S_\alpha(f)}(r)+\varsigma\log^+\frac{1}{|\Lambda_\alpha|}+\sum_{j=1}^{n}(r_j-r_0)m_{f,g_j}(r)+\\ &\quad \sum_{j=0}^{n}q_j m_{g_j,e_0}(r).\end{aligned}$$

By the First Main Theorem, 4.20, and assumption (B14) that each g_j grows more slowly than f,

$$\begin{aligned}\theta_\alpha(r) &\leq \hat{m}_{S_\alpha(f)}(r)+\sum_{j=1}^{n}(r_j-r_0)[T_f(r,s)+T_{g_j}(r,s)+m_{f,g_j}(s)]+\\ &\quad \sum_{j=0}^{n}q_j[T_{g_j}(r,s)+m_{g_j,e_0}(s)]+o(T_f(r,s))\\ &\underset{..}{\leq} \sum_{j=1}^{n}(r_j-r_0)T_f(r,s)+c_0(p)[\log^+ Ric_\tau(r,s)+\log^+ E(r)+\\ &\quad \log^+ Y_{\mathbb{B}}(r)]+o(T_f(r,s)),\end{aligned}$$

where $c_0(p)=7k(p)^2(k(p)+1)\varsigma$. q.e.d.

Now, define the function $\Lambda(r):\mathbb{R}_+\to\mathbb{R}$ by

$$\Lambda(r) = \int_{M<r>}\log\Xi\sigma. \tag{4.26}$$

LEMMA 4.7 *Suppose* [C1]-[C13] *are valid. For* $0<s<r$, $\Lambda(r)$ *satisfies the inequality*

$$\begin{aligned}\Lambda(r) &\underset{..}{\leq} \frac{k(p)[k(p)+1]}{2}Ric_\tau(r,s)+[q(p)+nq(p+1)]T_f(r,s)+\\ &\quad o(T_f(r,s)).\end{aligned} \tag{4.27}$$

PROOF: Recall from 4.4 and 4.6 the definitions of the functions K_λ and $\mathbb{K}$ such that for all λ in Λ,

$$\begin{aligned}K_\lambda &= (v_\lambda^0)^{q(p)}(v_\lambda^1\cdots v_\lambda^n)^{q(p+1)}\Delta_\lambda^{(p+1)[k(p)+1]};\\ \mathbb{K}|U_\lambda &= K_\lambda \overset{\triangle}{\mathfrak{v}}_\lambda^{\ k(p)+1}\otimes u_\lambda^{(p+1)[k(p)+1]}.\end{aligned}$$

Thus,

$$\begin{aligned}\mu_{\mathbb{K}}|U_\lambda &= \mu^0_{K_\lambda}\\ &= q(p)\mu^0_{v_\lambda^0}+q(p+1)\sum_{j=1}^{n}\mu^0_{v_\lambda^j}+(p+1)[k(p)+1]\mu^0_{\Delta_\lambda}.\end{aligned}$$

By Lemma 6.4 in Stoll [14], $N_\Delta(r,s,0) = o(T_f(r,s))$. Hence,

$$N_{\mathbb{K}}|U_\lambda \quad = \quad q(p)N_{f,a_0}(r,s) + q(p+1)\sum_{j=1}^{n} N_{f,a_j}(r,s) + o(T_f(r,s)).$$

Then, $\Lambda(r)$ can be bounded as follows:

$$\begin{aligned}
\Lambda(r) &= \int_{M<r>} \log \Xi\sigma \\
&= \int_{M<r>} \log \left(\frac{1}{\mathbf{Z}^{\frac{k(p)[k(p)+1]}{2}} |S(f)|} \prod_{j=0}^{n} \frac{1}{\lfloor f, a_j \rfloor^{q(p)}} \right) \sigma \\
&= -\int_{M<r>} \log|S(f)|\sigma - \int_{M<r>} \log \mathbf{Z}^{\frac{k(p)[k(p)+1]}{2}} \sigma + \\
&\quad \int_{M<r>} \sum_{j=0}^{n} \log \frac{1}{\lfloor f, a_j \rfloor^{q(p)}} \sigma \\
&= -\int_{M<r>} \log|S(f)|\sigma - \frac{k(p)[k(p)+1]}{2} \int_{M<r>} \log \mathbf{Z}\sigma + \\
&\quad \sum_{j=0}^{n} q(p) \int_{M<r>} \log \frac{1}{\lfloor f, a_j \rfloor} \sigma \\
&= -\int_{M<r>} \log|S(f)|\sigma - \frac{k(p)[k(p)+1]}{2} \int_{M<r>} \log \mathbf{Z}\sigma + \\
&\quad q(p)\sum_{j=0}^{n} m_{f,a_j}(r) \\
&= (-N_{S(f)}(r,s) - \int_{M<s>} \log|S(f)|\sigma) - \frac{k(p)[k(p)+1]}{2} \int_{M<r>} \log \mathbf{Z}\sigma + \\
&\quad q(p)\sum_{j=0}^{n} m_{f,a_j}(r) \qquad \text{[by the Jensen Formula for S(f)]} \\
&= -N_W(r,s) + N_{\mathbb{K}}(r,s) - \int_{M<s>} \log|S(f)|\sigma - \\
&\quad \frac{k(p)[k(p)+1]}{2} \int_{M<r>} \log \mathbf{Z}\sigma + q(p)\sum_{j=0}^{n} m_{f,a_j}(r) \ \ [\text{since } S(f) = \tfrac{W}{\mathbb{K}}] \\
&\le \frac{k(p)[k(p)+1]}{2} N_\beta(r,s) + N_{\mathbb{K}}(r,s) - \int_{M<s>} \log|S(f)|\sigma - \\
&\quad \frac{k(p)[k(p)+1]}{2} \int_{M<r>} \log \mathbf{Z}\sigma + q(p)\sum_{j=0}^{n} m_{f,a_j}(r) \qquad \text{[by 4.7]} \\
&= \frac{k(p)[k(p)+1]}{2} Ric_\tau(r,s) + N_{\mathbb{K}}(r,s) + q(p)\sum_{j=0}^{n} m_{f,a_j}(r) + \\
&\quad o(T_f(r,s)) \qquad \text{[by 1.6]}
\end{aligned}$$

$$
\begin{aligned}
&= \frac{k(p)[k(p)+1]}{2} Ric_\tau(r,s) + q(p)N_{f,a_0}(r,s) + q(p+1)\sum_{j=1}^{n} N_{f,a_j}(r,s) + \\
&\quad q(p)\sum_{j=0}^{n} m_{f,a_j}(r) + o(T_f(r,s)) \\
&\leq \frac{k(p)[k(p)+1]}{2} Ric_\tau(r,s) + q(p)[N_{f,a_0}(r,s) + m_{f,a_0}(r)] + \\
&\quad q(p+1)[\sum_{j=1}^{n} N_{f,a_j}(r,s) + m_{f,a_j}(r)] + o(T_f(r,s)) \\
&\underset{\cdot\cdot}{\leq} \frac{k(p)[k(p)+1]}{2} Ric_\tau(r,s) + [q(p) + nq(p+1)]T_f(r,s) + o(T_f(r,s)).
\end{aligned}
$$

q.e.d.

4.4 Product to sum estimates

The concepts of General Position and the Product to Sum Estimates will be introduced now and are further discussed in Chapter 4, Section 2 of Bardis [2]. As before, V is assumed to be a hermitian vector space of dimension $n+1$ and V^* is its dual vector space. For x_j in $\mathbb{P}(V^*)$ with $0 \leq j \leq n$, take $\mathfrak{x}_j$ in V_*^* such that $\mathbb{P}(\mathfrak{x}_j) = x_j$. Define

$$
\Box x_0 \wedge \ldots \wedge x_n \Box \quad = \quad \frac{\| \mathfrak{x}_0 \wedge \ldots \wedge \mathfrak{x}_n \|}{\| \mathfrak{x}_0 \| \cdots \| \mathfrak{x}_n \|}
$$

and therefore $0 \leq \Box x_0 \wedge \ldots \wedge x_n \Box \leq 1$, where $\Box x_0 \wedge \ldots \wedge x_n \Box = 0$ if and only if $x_0, \ldots, x_n$ are linearly dependent.

If A is contained in $\mathbb{P}(V^*)$ such that $n+1 \leq \#A < \infty$, then define

$$
\mathfrak{T}(A) \quad = \quad \{\alpha : \mathbb{Z}[0,n] \to A | \alpha \text{ is injective}\}.
$$

Also, define the gauge $\Gamma(A)$ to be

$$
\Gamma(A) \quad = \quad \min\{\Box \mu(0) \wedge \ldots \wedge \mu(n) \Box | \mu \in \mathfrak{T}(A)\}.
$$

Here, $\Gamma(A)$ is independent of the enumeration of A and $0 \leq \Gamma(A) \leq 1$. Then, $\Gamma(A) > 0$ if and only if A is in general position.

Consider a finite set $\mathfrak{G}$ of meromorphic maps $g : M \to \mathbb{P}(V^*)$ in general position with $\#\mathfrak{G} \geq n+1$. Let $\mathfrak{T} = \mathfrak{T}(\mathfrak{G})$ be the set of all injective maps $\alpha : \mathbb{Z}[0,n] \to \mathfrak{G}$. For each g in $\mathfrak{G}$, let $I(g)$ be the indeterminacy of g. Recall from 3.1 the definition of the indeterminacy of $\mathfrak{G}$, denoted $I(\mathfrak{G})$. In particular,

$$
I(\mathfrak{G}) \quad = \quad \bigcup_{g \in \mathfrak{G}} I(g).
$$

For every z in $M - I(\mathfrak{G})$, define the orbit by

$$
\mathfrak{G}(z) \quad = \quad \{g(z) | g \in \mathfrak{G}\}.
$$

Let $\epsilon_z : \mathfrak{G} \to \mathfrak{G}(z)$ be the surjective evaluation map given by $\epsilon_z(g) = g(z)$ for all g in $\mathfrak{G}$. The orbit $\mathfrak{G}(z)$ is called faithful if ϵ_z is bijective.

Then, the gauge of $\mathfrak{G}$, denoted $\Gamma(\mathfrak{G} : M - I(\mathfrak{G}) \to \mathbb{R}[0,1]$, is defined to be

$$\Gamma(\mathfrak{G})(z) \quad = \quad \min\{\square\mu(0)(z) \wedge \ldots \wedge \mu(n)(z)\square \,|\, \mu \in \mathfrak{T}\}$$

for z in $M - I(\mathfrak{G})$. Here, $0 \le \Gamma(\mathfrak{G})(z) \le 1$. Define the fusion $\hat{I}(\mathfrak{G})$ and the degeneracy $\tilde{I}(\mathfrak{G})$ by

$$\begin{aligned} \hat{I}(\mathfrak{G}) &= I(\mathfrak{G}) \cup \{z \in M - I(\mathfrak{G}) | \mathfrak{G}(z) \text{ is not faithful}\}; \\ \tilde{I}(\mathfrak{G}) &= I(\mathfrak{G}) \cup \{z \in M - I(\mathfrak{G}) | \Gamma(\mathfrak{G})(z) = 0\}. \end{aligned}$$

According to Lemma 7.13 and Lemma 7.14 of Stoll [12], both $\hat{I}(\mathfrak{G})$ and $\tilde{I}(\mathfrak{G})$ are analytic subsets of M such that

$$I(\mathfrak{G}) \quad \subseteq \hat{I}(\mathfrak{G}) \subseteq \tilde{I}(\mathfrak{G}) \not\equiv M.$$

Also, $\mathfrak{G}(z)$ is defined, faithful, and in general position if and only if z is in $M - \tilde{I}(\mathfrak{G})$.

Given these introductory remarks, the following result can now be stated.

THEOREM 4.8 [PRODUCT TO SUM ESTIMATES] *Take* λ *in* $\mathbb{Z}_+$. *Suppose* [C1]-[C2] *and* [C9]-[C10] *are valid. Let* $a = \#\mathfrak{G} \ge n + \lambda$. *Let* $\mathfrak{K}_\lambda$ *be the set of all subsets* N *of* $\mathfrak{G}$ *with* $\#N = n + \lambda$. *Let* $P : \mathbb{P}(V) \times M \times \mathfrak{G} \to \mathbb{R}[0,1]$ *be a function. Choose* x *in* $\mathbb{P}(V)$ *and* z *in* $M - \tilde{I}(\mathfrak{G})$ *such that* $\square x, g(z)\square > 0$ *for all* g *in* $\mathfrak{G}$. *Then,*

$$\prod_{g \in \mathfrak{G}} \frac{P(x,z,g)}{\square x, g(z)\square} \quad \le \quad \left[\frac{2^{3n+3}}{\Gamma(\mathfrak{G})}\right]^{a-n-\lambda} \sum_{N \in \mathfrak{K}_\lambda} \left[\frac{P(x,z,g)}{\square x, g(z)\square}\right]. \tag{4.28}$$

PROOF: See Theorem 6.2 of Ru-Stoll [7].

Let $\mathfrak{T}$ be given as before and $g_j = \alpha(j)$. If $\mathfrak{G}$ is in general position, define the gauge measure function $\Gamma_\mathfrak{G} : \mathfrak{C}_\tau \to \mathbb{R}_+$ by

$$\Gamma_\mathfrak{G}(r) \quad = \quad \int_{M<r>} \log \frac{1}{\Gamma(\mathfrak{G})} \sigma. \tag{4.29}$$

This integral exists by Theorem 7.15 of Stoll [12]. It gives some measure of the decline in general position as $r \to \infty$. By Theorem 7.16 of Stoll [12], if $q > n$ and $r > s > 0$, then

$$\Gamma_\mathfrak{G}(r) \quad \le \quad \binom{q-1}{n} \sum_{g \in \mathfrak{G}} T_g(r,s) + m_\mathfrak{G}(s),$$

where $m_\mathfrak{G}(s)$ is constant. Hence, as $r \to \infty$,

$$\Gamma_G(r) \quad \le \quad o(T_f(r,s)). \tag{4.30}$$

4.5 Moving targets and the defect relation

Now, everything is ready to state and prove the culmination of this work, the Defect Relation for Moving Targets. First, the following important result must be derived.

THEOREM 4.9 [THE SECOND MAIN THEOREM FOR MOVING TARGETS] *Suppose* [C1]-[C14] *are valid. Take* $0 < s < r$. *Define* $a = \#\mathfrak{G} > n + 1$ *and* $b = \#\mathfrak{T}$. *Then, for every* $\epsilon > 0$, *there exists* $p = p(\epsilon) > s(n)$ *such that*

$$\begin{aligned}\sum_{g\in\mathfrak{G}} m_{f,g}(r) &\underset{..}{\leq} (1+n+\epsilon)T_f(r,s) + \frac{k(p)[k(p)+1]}{2q(p)}Ric_\tau(r,s) + \\ &\quad \frac{bc_0(p)}{q(p)}[\log^+ Ric_\tau(r,s) + \log^+ E(r) + \log^+ Y_{\mathbb{B}}(r)]. \qquad (4.31)\end{aligned}$$

PROOF: By the Product to Sum Estimates 4.28 and the fact that there exists $(n+1)!$ permutations of a set of $n+1$ elements,

$$\begin{aligned}\prod_{g\in\mathfrak{G}} \frac{1}{\square f, g\square} &\leq \frac{1}{(n+1)!}\left[\frac{2^{3n+3}}{\Gamma(\mathfrak{G})}\right]^{a-n-1} \sum_{\alpha\in\mathfrak{T}} Q_\alpha \\ &= \frac{1}{(n+1)!}\left[\frac{2^{3n+3}}{\Gamma(\mathfrak{G})}\right]^{a-n-1} \Xi^{\frac{1}{q(p)}} \sum_{\alpha\in\mathfrak{T}} (R_\alpha)^{\frac{1}{q(p)}}.\end{aligned}$$

This then implies that

$$\begin{aligned}\sum_{g\in\mathfrak{G}} \log\frac{1}{\square f, g\square} &\leq (a-n-1)\log\frac{1}{\Gamma(\mathfrak{G})} + \frac{1}{q(p)}\log\Xi + \log\sum_{\alpha\in\mathfrak{T}}(R_\alpha)^{\frac{1}{q(p)}} + \\ &\quad (a-n-1)(3n+3)\log 2 \\ &\leq (a-n-1)\log\frac{1}{\Gamma(\mathfrak{G})} + \frac{1}{q(p)}\log\Xi + \frac{1}{q(p)}\sum_{\alpha\in\mathfrak{T}}\log^+ R_\alpha + c,\end{aligned}$$

where $c = (a-n-1)(3n+3)\log 2 + \log b$. Then, by 4.24, 4.26, and 4.29, integrating over $M < r >$ for $0 < s < r$ yields

$$\sum_{g\in\mathfrak{G}} m_{f,g}(r) \leq (a-n-1)\Gamma_G(r) + \frac{1}{q(p)}\Lambda(r) + \frac{1}{q(p)}\sum_{\alpha\in\mathfrak{T}}\theta_\alpha(r) + c\varsigma.$$

Then, by 4.25, 4.27, and 4.30,

$$\begin{aligned}\sum_{g\in\mathfrak{G}} m_{f,g}(r) &\leq \frac{1}{q(p)}\Lambda(r) + \frac{1}{q(p)}\sum_{\alpha\in\mathfrak{T}}\theta_\alpha(r) + o(T_f(r,s)) \\ &\underset{..}{\leq} \left[\frac{q(p)+nq(p+1)}{q(p)}\right]T_f(r,s) + \frac{k(p)[k(p)+1]}{2q(p)}Ric_\tau(r,s) + \\ &\quad \sum_{j=1}^{n}\frac{b(r_j - r_0)}{q(p)}T_f(r,s) + \frac{bc_0(p)}{q(p)}[\log^+ Ric_\tau(r,s) + \\ &\quad \log^+ E(r) + \log^+ Y_{\mathbb{B}}(r)] + o(T_f(r,s))\end{aligned}$$

$$= \left[1 + n\frac{q(p+1)}{q(p)}\right] T_f(r,s) + \sum_{j=1}^{n} \frac{b(r_j - r_0)}{q(p)} T_f(r,s) + \frac{k(p)[k(p)+1]}{2q(p)} Ric_\tau(r,s) + \frac{bc_0(p)}{q(p)}[\log^+ Ric_\tau(r,s) + \log^+ E(r) + \log^+ Y_{\mathbb{B}}(r)] + o(T_f(r,s)).$$

Now, by Lemma 7.2 of Ru-Stoll [7], for any $\epsilon > 0$, there exists a positive integer $p = p(\epsilon) > s(n)$ such that

$$1 \leq \frac{q(p+1)}{q(p-s(j))} < 1 + \frac{\epsilon}{3n} \text{ for } 0 \leq j \leq n;$$

$$1 \leq \frac{q(p-s(i))}{q(p-s(j))} < 1 + \frac{\epsilon}{3n} \text{ for } 0 \leq i \leq j \leq n;$$

$$0 \leq \frac{r_j(p)-r_0(p)}{q(p)} < \frac{\epsilon}{3nb} \text{ for } 0 \leq j \leq n.$$

Thus,

$$\sum_{g\in\mathfrak{G}} m_{f,g}(r) \underset{\because}{\leq} \left[1 + n\left(1+\frac{\epsilon}{3n}\right)\right] T_f(r,s) + \sum_{j=1}^{n} b\left(\frac{\epsilon}{3nb}\right) T_f(r,s) + \frac{k(p)[k(p)+1]}{2q(p)} Ric_\tau(r,s) + \frac{bc_0(p)}{q(p)}[\log^+ Ric_\tau(r,s) + \log^+ E(r) + \log^+ Y_{\mathbb{B}}(r)] + o(T_f(r,s))$$

$$= \left[1 + n + \frac{\epsilon}{3}\right] T_f(r,s) + bn\left(\frac{\epsilon}{3nb}\right) T_f(r,s) + \frac{k(p)[k(p)+1]}{2q(p)} Ric_\tau(r,s) + \frac{bc_0(p)}{q(p)}[\log^+ Ric_\tau(r,s) + \log^+ E(r) + \log^+ Y_{\mathbb{B}}(r)] + \frac{\epsilon}{3} T_f(r,s)$$

$$= [1+n+\epsilon] T_f(r,s) + \frac{k(p)[k(p)+1]}{2q(p)} Ric_\tau(r,s) + \frac{bc_0(p)}{q(p)}[\log^+ Ric_\tau(r,s) + \log^+ E(r) + \log Y_{\mathbb{B}}(r)].$$

q.e.d.

THEOREM 4.10 [THE DEFECT RELATION] *Suppose* [C1]-[C14] *are valid. Also, assume that* $\limsup_{r\to\infty} \frac{Ric_\tau(r,s)}{T_f(r,s)} = 0$; $\limsup_{r\to\infty} \frac{\log^+ E(r)}{T_f(r,s)} = 0$; *and* $\limsup_{r\to\infty} \frac{\log^+ Y_{\mathbb{B}}(r)}{T_f(r,s)} = 0$. *Then,*

$$\sum_{g\in\mathfrak{G}} \delta(f,g) \leq n+1.$$

PROOF: Recall that

$$\sum_{g\in\mathfrak{G}} \delta(f,g) = \sum_{g\in\mathfrak{G}} \liminf_{r\to\infty} \frac{m_{f,g}(r)}{T_f(r,s)+T_g(r,s)}.$$

Then, simplifying and using 4.31,

$$\begin{aligned}
\sum_{g\in\mathfrak{G}} \delta(f,g) &= \sum_{g\in\mathfrak{G}} \liminf_{r\to\infty} \frac{m_{f,g}(r)}{T_f(r,s)} \\
&\leq \liminf_{r\to\infty} \sum_{g\in\mathfrak{G}} \frac{m_{f,g}(r)}{T_f(r,s)} \\
&\leq \limsup_{r\to\infty} \left\{ \frac{(1+n+\epsilon)T_f(r,s)}{T_f(r,s)} + \frac{\frac{k(p)[k(p)+1]}{2q(p)} Ric_\tau(r,s)}{T_f(r,s)} + \right. \\
&\qquad \left. \frac{\frac{bc_0(p)}{q(p)}[\log^+ Ric_\tau(r,s) + \log^+ E(r) + \log^+ Y_{\mathbb{B}}(r)]}{T_f(r,s)} \right\} \\
&= 1+n+\epsilon.
\end{aligned}$$

Therefore, this estimate is valid for every $\epsilon > 0$. Consequently, ϵ need not be kept fixed, but instead ϵ may now converge to zero. This yields

$$\sum_{g\in\mathfrak{G}} \delta(f,g) \leq n+1.$$

q.e.d.

Bibliography

[1] L. Ahlfors, The Theory of Meromorphic Curves, Acta Soc. Sci. Fenn, Nova Ser. A 3, 4 (1941), 171-183.

[2] E. Bardis, The Defect Relation for Meromorphic Maps Defined on Covering Parabolic Manifolds, Thesis, Notre Dame (1990).

[3] H. Cartan, Sur les Zeros des Combinaisons Lineaires de p Fonctions Holomorphes Donnees, Mathematica (cluj) 7 (1933), 80-103.

[4] C. T. Chuang, Une Generalisation d'une Inegalite de Nevanlinna, Scientia Sinica, 13 (1964), 887-895.

[5] R. Nevanlinna, Le Theoreme de Picard-Borel et la Theorie des Fonctions Meromorphes, Gauthiers-Villars, Paris (1929), reprint Chelsea-Publ. Co., New York (1974).

[6] M. Ru-W. Stoll, The Nevanlinna Conjecture for Moving Targets, Geometrical and Algebraic Aspects in Several Complex Variables (June 1989), 293-308. Cetraro (Italy).

[7] M. Ru-W. Stoll, The Second Main Theorem for Moving Targets, The Journal of Geometric Analysis, 1, 2 (1991), 99-138.

[8] N. Steinmetz, Eine Verallgemeinerung de zweiten Nevanlinnaschen Hauptsatzes, J.Reine Angew. Math, 368 (1986), 134-141.

[9] W. Stoll, Deficit and Bezout Estimates, Value Distribution Theory Part B (Edited by R.O. Kujula and A.L. Vitter III), Pure and Applied Math., 25 (1973), 1-271. Marcel Dekker.

[10] W. Stoll, Value Distribution on Parabolic Spaces, Lecture Notes in Mathematics, 600 (1977). Springer Verlag.

[11] W. Stoll, Introduction to Value Distribution of Meromorphic Maps, Lecture Notes in Mathematics, 950 (1982), 210-359. Springer Verlag.

[12] W. Stoll, Value Distribution Theory for Meromorphic Maps, Aspekte der Mathematik, E7 (1985). Friedr. Vieweg & Sohn.

[13] W. Stoll, Algebroid Reduction of Nevanlinna Theory, Lecture Notes in Mathematics, 1277 (1987), 131-241. Springer Verlag.

[14] W. Stoll, An Extension of the Theorem of Steinmetz-Nevanlinna to Holomorphic Curves, Math. Ann., 282 (1988), 185-222.

[15] W. Stoll, Value Distribution Theory in Several Complex Variables (1994). Shandong Press.

[16] Ch. Tung, The First Main Theorem of Value Distribution on Complex Spaces, Atti della Acc. Naz. d. Lincei Serie VIII, 15 (1979), 93-262.

[17] A. Vitter, The Lemma of the Logarithmic Derivative in Several Complex Variables, Duke Math. J., 44 (1977), 89-104.

[18] H. Weyl and J. Weyl, Meromorphic Functions and Analytic Curves, Annals of Mathematics Studies, 12 (1943). Princeton University Press.

Editorial Information

To be published in the *Memoirs*, a paper must be correct, new, nontrivial, and significant. Further, it must be well written and of interest to a substantial number of mathematicians. Piecemeal results, such as an inconclusive step toward an unproved major theorem or a minor variation on a known result, are in general not acceptable for publication. *Transactions* Editors shall solicit and encourage publication of worthy papers. Papers appearing in *Memoirs* are generally longer than those appearing in *Transactions* with which it shares an editorial committee.

As of January 31, 1999, the backlog for this journal was approximately 4 volumes. This estimate is the result of dividing the number of manuscripts for this journal in the Providence office that have not yet gone to the printer on the above date by the average number of monographs per volume over the previous twelve months, reduced by the number of issues published in four months (the time necessary for preparing an issue for the printer). (There are 6 volumes per year, each containing at least 4 numbers.)

A Copyright Transfer Agreement is required before a paper will be published in this journal. By submitting a paper to this journal, authors certify that the manuscript has not been submitted to nor is it under consideration for publication by another journal, conference proceedings, or similar publication.

Information for Authors and Editors

Memoirs are printed by photo-offset from camera copy fully prepared by the author. This means that the finished book will look exactly like the copy submitted.

The paper must contain a *descriptive title* and an *abstract* that summarizes the article in language suitable for workers in the general field (algebra, analysis, etc.). The *descriptive title* should be short, but informative; useless or vague phrases such as "some remarks about" or "concerning" should be avoided. The *abstract* should be at least one complete sentence, and at most 300 words. Included with the footnotes to the paper, there should be the 1991 *Mathematics Subject Classification* representing the primary and secondary subjects of the article. This may be followed by a list of *key words and phrases* describing the subject matter of the article and taken from it. A list of the numbers may be found in the annual index of *Mathematical Reviews*, published with the December issue starting in 1990, as well as from the electronic service e-MATH [**telnet e-MATH.ams.org** (or **telnet 130.44.1.100**). Login and password are **e-math**]. For journal abbreviations used in bibliographies, see the list of serials in the latest *Mathematical Reviews* annual index. When the manuscript is submitted, authors should supply the editor with electronic addresses if available. These will be printed after the postal address at the end of each article.

Electronically prepared papers. The AMS encourages submission of electronically prepared papers in AMS-TeX or AMS-LaTeX. The Society has prepared author packages for each AMS publication. Author packages include instructions for preparing electronic papers, the *AMS Author Handbook*, samples, and a style file that generates the particular design specifications of that publication series for both AMS-TeX and AMS-LaTeX.

Authors with FTP access may retrieve an author package from the Society's Internet node `e-MATH.ams.org` (130.44.1.100). For those without FTP access, the author package can be obtained free of charge by sending e-mail to `pub@ams.org` (Internet) or from the Publication Division, American Mathematical Society, P.O. Box 6248, Providence, RI 02940-6248. When requesting an author package, please specify $\mathcal{A}\mathcal{M}\mathcal{S}$-TeX or $\mathcal{A}\mathcal{M}\mathcal{S}$-LaTeX, Macintosh or IBM (3.5) format, and the publication in which your paper will appear. Please be sure to include your complete mailing address.

Submission of electronic files. At the time of submission, the source file(s) should be sent to the Providence office (this includes any TeX source file, any graphics files, and the DVI or PostScript file).

Before sending the source file, be sure you have proofread your paper carefully. The files you send must be the EXACT files used to generate the proof copy that was accepted for publication. For all publications, authors are required to send a printed copy of their paper, which exactly matches the copy approved for publication, along with any graphics that will appear in the paper.

TeX files may be submitted by email, FTP, or on diskette. The DVI file(s) and PostScript files should be submitted only by FTP or on diskette unless they are encoded properly to submit through e-mail. (DVI files are binary and PostScript files tend to be very large.)

Files sent by electronic mail should be addressed to the Internet address `pub-submit@ams.org`. The subject line of the message should include the publication code to identify it as a Memoir. TeX source files, DVI files, and PostScript files can be transferred over the Internet by FTP to the Internet node `e-math.ams.org` (130.44.1.100).

Electronic graphics. Figures may be submitted to the AMS in an electronic format. The AMS recommends that graphics created electronically be saved in Encapsulated PostScript (EPS) format. This includes graphics originated via a graphics application as well as scanned photographs or other computer-generated images.

If the graphics package used does not support EPS output, the graphics file should be saved in one of the standard graphics formats—such as TIFF, PICT, GIF, etc.—rather than in an application-dependent format. Graphics files submitted in an application-dependent format are not likely to be used. No matter what method was used to produce the graphic, it is necessary to provide a paper copy to the AMS.

Authors using graphics packages for the creation of electronic art should also avoid the use of any lines thinner than 0.5 points in width. Many graphics packages allow the user to specify a "hairline" for a very thin line. Hairlines often look acceptable when proofed on a typical laser printer. However, when produced on a high-resolution laser imagesetter, hairlines become nearly invisible and will be lost entirely in the final printing process.

Screens should be set to values between 15% and 85%. Screens which fall outside of this range are too light or too dark to print correctly.

Any inquiries concerning a paper that has been accepted for publication should be sent directly to the Editorial Department, American Mathematical Society, P. O. Box 6248, Providence, RI 02940-6248.

Editors

Selected Titles in This Series

(Continued from the front of this publication)

(See the AMS catalog for earlier titles)